DATE DUE

MY 05			

DEMCO 38-296

Best Practice in Inventory Management

Best Practice in Inventory Management

Tony Wild

JOHN WILEY & SONS, INC.

New York • Chichester • Weinheim • Brisbane • Singapore • Toronto

This book is printed on acid-free paper. ∞

This publication is designed to provide accurate and authoritative information in regard to
the subject matter covered. It is sold with the understanding that the publisher is not engaged
in rendering legal, accounting, or other professional services. If legal advice or other expert
assistance is required, the services of a competent professional person should be sought.

ISBN 0-471-25341-3

Printed in the United States of America.

10 9 8 7 6 5 4 3 2

Contents

Preface

In the distant past when I ascended from a being a scientist to the heights of inventory analysis, I was surprised at the lack of helpful written work on the subject. Although there are now many books, manuals and papers available, there does not seem to be much written on the very important subject of how to carry out inventory control faced with the real inventory, customers and an assortment of stock records and styles of suppliers. Yet better control of stock (inventory) can give major benefits to the profitability of all companies. This book is aimed to show how good inventory control can be used in practice. It is a result of working continuously on inventory control with a large number of companies over many years. It contains the distilled techniques which have been tried out and proved to work.

This book contains two main ingredients, namely:

- The basics of inventory management as covered in the IOM (Institute of Operations Management) Diploma syllabus.
- The application of these techniques to real inventory management.

The applications are what really count, since the knowledge is not at all interesting without application. The experience of the author has been to use the basics to provide powerful changes in stocks and profitability. In several cases millions of pounds of inventory value have been saved and in others customer service has been greatly improved. Many of the topics here have been tried, tested and approved. They are all a matter of common sense, although which piece of common sense to use in a particular situation is a matter which needs deeper understanding.

The book is a driver's manual for inventory controllers. It will cover the working of the engine (how inventory control techniques work), how to use the controls (what the techniques do and how to manage them) and how to get the best out of the vehicle (how to optimise inventory). Understand the text and it will show the way to guaranteed improved inventory control, reduced stock levels and higher availability.

In the preparation of this book there have been contributions from the many thousands of people to whom I have presented courses on these ideas and who have helped to hone the concepts. Thanks are also due to those companies who have let me loose on their inventory, carrying out consultancy projects which have extended my knowledge and proved the effectiveness and scope of different techniques.

Central to the development of inventory techniques has been the Institute of Operations Management (formerly BPICS – the British Production and Inventory Control Society) which has been a channel for professional debate, friendship and technical information. This book has been created with the support of Richard Turner as secretary of the Institute.

The evolution of this book has been a two handed operation. The concepts have been blended into a logical text through my own work and that of Elaine Duckworth, who has shared the task of developing a useful and readable handbook.

DR TONY WILD

Introduction

Everyone is a stock controller, at home and at work. We all keep food, clothes, domestic items, paper, pens and many other goods. We also have shortages and emergency purchases. Some people regularly have to throw out the contents of the refrigerator because they have been there for a while and changed in character. So stock control is a natural occupation which everyone does, some more successfully than others.

There are many other activities which people do, such as sport, carpentry, music and medicine, and for these it is generally accepted that a level of technical expertise is needed. In carpentry and music there are people who become highly skilled through their natural talents. Success appears to come from a mixture of talent, determination, practice and knowledge. In medicine, particularly in surgery, we are more likely to rely on those who have studied the subject and have a deep knowledge of both theory and practice. Lawyers cannot function without knowledge; often a common-sense approach does not work for them.

As with any profession, stock control (or the subject of professional inventory management) has a body of techniques and knowledge which differentiates the professional from the DIY enthusiast. This expertise is the result of 100 years of development and refining. Now there are professional inventory managers with recognised qualifications, including the Diploma and Certificate in Professional Inventory Management (DPIM and CPIM, respectively). There are still colleagues who believe that inventory can be managed using a clear brain and common sense. It is unlikely that they would be happy with the same competency in their doctor or financial adviser.

This book is for the inventory control practitioner. Many people have been able to manage their stock of inventory with the techniques described so that the customers are happier and so are the accountants. Reduction in inventory value, avoidance of unnecessary work and improvement in customer service can be accomplished at the same time through simple application of these techniques. Improved inventory management has been shown to halve stocks and improve service at the same time.

The techniques described are the basic concepts which inventory controllers should have at their fingertips. The more complex theoretical approaches can be left to those developing sophisticated systems or those with excellent inventory control (and over the years I have not met any of these!). Inventory practitioners should be able to use this book to understand the best approaches and then to apply them to their own circumstances. Simple application of the methods is most successful, while modifications usually result in less effective outcomes.

The text covers the syllabus for the technical qualifications including DPIM and CPIM. It discusses good practice, inventory theory and the practical application of the techniques. The book starts with inventory structure, explaining target setting in chapters 1 and 2, and how to structure inventory in chapter 3. The chapter shows how easily to make a big impact on inventory levels and the approach gives major benefits in tightening control. Continuing the theme of restructuring stock leads to the ultimate in stock management, the 'zero inventory' approach, 'Just In Time' (JIT), discussed in chapter 4. Gradual absorption of the philosophy throughout inventory management is proving a great benefit and it is worth examining how to utilise JIT in each stock environment. The rate of progress is often governed by the attitude and organisation in a company and the development of modern inventory management practice is reviewed in chapter 5.

The discussion follows on to the detail of item level inventory management, the essence of real inventory management. Chapters 6 to 8 are the elixir of item level stock control, the secret recipe for success. Chapter 6 concentrates on how to keep out of trouble: a rational approach to the risks of stockouts, chapter 7 shows how to use this information to set stock levels and chapter 8 discusses the options for ensuring suppliers are agreeable and flexible.

The third aspect of inventory management, one which is usually carried out poorly, is forecasting. Forecasting methods are relatively well advanced, but there is less practical application of these in most companies. This is an opportunity, discussed in chapter 9, while the basic techniques are discussed in chapter 10. Chapter 10 also includes the techniques used to ensure that forecasts produce sensible results. In practice there are items where it is worth managing the stock in more detail, so chapter 11 gives more techniques that are commonly used and provide good forecasts.

The 'one item at a time' approach to inventory management is really not appropriate in many situations. Users often require a range of items to do a job. For example, servicing the car requires that filters, plugs, oil, and so on, all be available at one time. Inventory controllers should therefore balance the stocks to meet the structure demand. Unfortunately the usual techniques do not help and so the approach described in chapter 12 is applied. This technique, developed over many years in manufacturing, is an important tool for general inventory management, although the application does not need the complexity which is used in manufacturing.

The inventory manager who has a good knowledge of these topics can then look round for new challenges. Industry has not been slow to provide these. In addition to the increasing pressure on inventory value and availability levels, new dimensions are being added in the form of logistics and supply chain management. If the balancing of inventory in one store is not sufficient, the inventory, distribution costs and manufacturing efficiency have all to be optimised at the same time. Chapter 13 reviews the opportunities in this area.

Chapters 3, 6 and 7 are the bedrock of inventory practice. From experience, inventory management as described in this book works well and the purpose of these pages is to enable more individuals to use the methods with confidence.

Readers outside the UK may find that a little translation is required. For instance in the USA the term 'inventory management' is applied to the whole subject, whereas in the UK it is normally known as 'stock control'. The terms 'inventory' and 'stock' are both used in the text to mean the same (in the UK inventory has a financial flavour and stock an operational context). It is hoped that this and other nuances of translation do not cloud the understanding of this technical and fascinating subject, whatever it is called.

1

The basis of inventory control

Role of inventory management

The success of a venture depends on its ability to provide services to customers or users and remain financially viable. For an organisation which is supplying goods to its customers, the major activity is to have suitable products available at an acceptable price within a reasonable timescale. Many parts of a business are involved in setting up this situation. Initially it is the marketing and design departments. Then purchasing, and in some cases, manufacture are involved. For an item which is already in the marketplace, the main activity is providing a continuity of supply for the customers.

Inventory control is the activity which organises the availability of items to the customers. It coordinates the purchasing, manufacturing and distribution functions to meet marketing needs. This role includes the supply of current sales items, new products, consumables, spare parts, obsolescent items and all other supplies. Inventory enables a company to support its customer service, logistic or manufacturing activities in situations where purchase or manufacture of the items is not able to satisfy demand. Lack of satisfaction could arise either because the speed of purchasing or manufacturing is too protracted, or because the appropriate quantities cannot be provided without stocks.

Stock control[1] exists at a crossroads in the activities of a company. Many of the activities depend on the correct level of stock being held, but the definition of the term 'correct level' varies depending upon which activity is defining the stock. Stock control is definitely a balancing act between the conflicting requirements of the company, and the prime reason for the development of inventory management is to resolve this conflict in the best interest of the business. A conventional supply organisation will have many departments including sales, purchasing, finance, quality assurance,

---

[1] Note the terms 'inventory' and 'stock' are used to mean the same thing in this book.

contracts and general administration. In some cases there will also be manufacturing, distribution or support services or a variety of industry specific activities. Each of these has a particular view of the role of stock control.

Sales consider that good stock control enables the company to have available any item which will meet immediate sales for as large a quantity as demanded: this requires large stock. For service companies where parts service is involved, the control of stock at the customer interface is traditionally left with the person carrying out the service, and this has led to overstocking and poor control. Similarly in distribution, the effect of bulking up shipments will lead to high stock levels and a compromise has to be reached.

Purchasing consider that stock control provides the opportunity for goods to be purchased so that optimum prices can be obtained. Buying items in bulk often reduces the purchase price and it also improves the efficiency within the purchasing department. The stores is a means of keeping the bulk purchase items after buying advantageously.

Finance departments have a problem with stock because it consumes vast amounts of working capital and upsets the cash flow. One benefit of stock from a financial standpoint is that provisions can be made in case the stock turns out to be unsaleable, and this value can be adjusted to modify the profit figure in times of good or bad financial results. However the existence of these provisions in the first place is detrimental to the finances of the company.

Quality management normally has the effect of slowing down the progress of stock while the necessary checks are made. This means that quality and inventory personnel effectively work in opposition. The gradual introduction of formal quality standards for supply and manufacture has reduced this conflict in most organisations. Supplier conformance has been improved enabling intrusive checking to be minimised.

General management see stock control, rightly, as a source of information. Some management considers that they should be able to use stock control to give an immediate supply of information, statistics and forecasts. This can result in large amounts of unstructured work in collecting, analysing and providing information.

The traditional view taken by manufacturing companies has been that large batches reduce direct production costs. Manufacturing management tends to aim more for plant and labour efficiency and allow high stocks in order to avoid the disruptions caused by shortages, breakdowns and changing customer demand. Good stock controllers (or materials management) will keep the stock down as long as they are responsible for stocks of everything, including raw materials, finished and semi-finished stocks, consumables, tools, work on the shop floor and, in fact, all inventory items.

These views are now subject to question. Those who agree with these ideas should be considering a rethink and change to a consensus on stock management. The development of stock control practice has been given impetus by the move toward total quality management and just in time concepts. These have focused on the possibility that good management, particularly, is both desirable and profitable, and has brought into question all the attitudes which are presented above as alternative views of stock control.

The responsibility for maintaining the correct balance is normally left to inventory management. (The actual name for the activity varies but the task remains the same: the controller could be materials or inventory manager, stock planner or controller, or even considered as a buyer or logistics controller.)

Demand for the products results from the changes in market and financial forces and the amount and type of stock control varies in tune with this. Stock control is a dynamic activity and the successful inventory manager has to ensure that the balance is kept right. This requires both communications skills and professional inventory techniques. Operating methods should be continuously revised to reflect the changes and systems should be altered to suit new situations and operating policies.

Objectives for inventory control

The first consideration is the overall objective of the work of stock control. Like all other activities in the company, inventory management has to contribute to the welfare of the whole organisation. The logistic operation must aim to 'contribute to profit by servicing the marketing and financial needs of the company.' The aim is not to make all items available at all times as this may well be detrimental to the finances of the company. The normal role for stock control is to 'meet the required demand at a minimum cost.'

The aim of long term profitability has to be translated into operational and financial targets which can be applied to daily operations. The purpose of the inventory control function in supporting business activities is to optimise three targets:

- Customer service
- Inventory costs
- Operating costs.

The most profitable policy is not to optimise one of these at the expense of the others. The inventory controller has to make value judgements. If profit is lacking the company goes out of business in the short term. If customer service is poor, then the customers disappear and the company goes out of business in the longer term. Balancing the financial and marketing aspects is the answer: the stock controller has a fine judgement to make.

The first target, customer service, can be considered in several ways, depending on the type of demand (discussed in chapter 2). In a general stores environment the service will normally be taken as 'availability ex stock', whereas in a supply to customer specification, the service expected would be delivery on time against customer requested date.

The second target, inventory costs, requires a minimum of cash tied up in stock. This target has to be considered carefully, since there is often the feeling that having any stock in stores for a few months is bad practice. In reality, minimising the stock usually means attending to the major costs: very low value items are not considered a significant problem. Low inventory can also be considered in terms of space or other critical resources. Where the item is voluminous or the stores space restricted, the size of the items will also be a major consideration.

The third target, avoiding operating costs, has become more of an issue as focus has been placed on stock management. The prime operating costs are the stores operations, stock control, purchasing and associated services. The development of logistics, linking distribution costs with inventory, has added a new set of transportation costs to the analysis.

Optimising the balance between these three objectives is the focus of stock control. The better the balance the greater the profits provided for the company (as will be shown in the section on *Profit through inventory management*). The improvement in stock control has been slow and gradual, created by new technology, financial need and competitive pressure. Those companies who can tighten their control faster than the average will flourish, but those which do not keep up with the average, even if they are improving, will gradually dwindle. The way to achieve much is relatively simple and is outlined in the remainder of this book. The trick of the good stock controller is to meet the objectives simultaneously, not one at a time, and of course 'the better the control the smaller the cost, the lower the stock levels, and the better the customer service'.

One of the dichotomies of inventory control is that at item level, the more stock the better the availability (this is discussed quantitatively in chapter 6 in the section on *Evaluation of safety stocks*). However for the whole inventory, experience has shown that the businesses with the highest stock are often those which have the worst availability. These observations are not in conflict if the causes are considered. Stockouts result from holding too little stock of the offending lines, because the forecasts, monitoring or controls are inadequate. High stock levels arise because too much stock has been purchased through bad forecasting, monitoring or controls. High stock and poor availability are caused simultaneously as a result of poor control. The problem rests with the inventory controller and the solution is in improved techniques. By using the techniques discussed in the remainder of this book the reader can reduce stockholding by a quarter even where the control is already reasonable.

Profit through inventory management

An example of the effect of stock on profitability

There are two companies in the same industry, M Tight Ltd. and The Slack Company. Each has annual sales of $5 million and employs 85 people. They both have fixed assets of $2½ million in plant and buildings and operate in an industry where the value of customer debtors is balanced by the credit owed to suppliers.

The two companies differ in their organisation. Tight is a small, independent company and Slack is a small part of a multinational group. Although they manage to achieve similar customer service, M Tight Ltd. has concentrated on inventory control and has reduced stocks to $½ million, while The Slack Company have poorly defined responsibilities in this area and hold $2½ million worth of stock.

The Slack Company have to finance an extra $2 million inventory and they can borrow this from their Head Office. However, the company expects a 15% return on investment and charges Slack Company $300 000 pa in interest. They also have the costs of controlling, holding and organising the extra stock which is stored in an adjacent warehouse. These warehousing costs are as follows:

	Cost ($ pa)
Stores: extra area rent, heating etc.	500
Stores, equipment; trucks and racking	
depreciation	1 000
maintenance	1 000
Stock obsolescence	5 000

	Cost ($ pa)
Two storekeepers' wages and employment costs	12 000
Annual stocktaking	500
Control systems and computer time	3 000
TOTAL	23 000
Plus financing charge $2 million @ 15%	300 000
TOTAL EXTRA COST	$323 000

Now examine the profit and loss accounts for the two companies.

M Tight Ltd. have a return on turnover of 5%, making $250 000 profit on the $5 m sales. The Slack has extra costs of $323 000 and thus makes a loss of $73 000. Good stock control has made the difference between profit for M Tight Ltd. and loss for The Slack Company.

The next year M Tight Ltd. made $500 000 profit on the same $5 million turnover, a 'return on sales' of 10%. The Slack Company had a profit of $500 000 − $323 000 = $177 000 because of their annual stock-holding costs.

Although M Tight Ltd. has a return on sales of 10%, The Slack Company only obtain

$$\frac{177000}{5000000} = 3\tfrac{1}{2}\%.$$

This shows how important it is to the profit of a company that stockholding costs are minimised. It is a surprisingly significant difference, and when we look at return on capital, the situation is even more dramatic.

Both companies have plant and buildings worth $2\tfrac{1}{2}$ million, with total assets consisting of the fixed element of plant and machinery, working capital tied up in inventory and creditors who are offset by outstanding debtors. M Tight Ltd. also has $\tfrac{1}{2}$ million worth of stock, making a total capital investment of $3 million. The return on assets for M Tight Ltd. is therefore

$$\frac{500000}{3000000} = 16\tfrac{2}{3}\%.$$

The Slack Company has $2\tfrac{1}{2}$ million of stock making a total investment of $5 million. Their return on assets is therefore

$$\frac{177000}{5000000} = 3^1/_2\%.$$

By running the same business with two different inventory plans, anything from 17% to 3% can be earnt for the trouble taken.

M Tight Ltd. has cash available for investment, wages, better working conditions and to share out to the pension schemes, insurance companies and other investors in their company. The Slack Company have little cash to spare other than to pay off the bank. Imagine the situation, poor profitability leading to disillusionment, poor working conditions and fears for the future.

Reasons for the current stock

It is easy to assess the effectiveness of inventory control in a company. All that is required is a visit to the stores. The stock level is a result of the effectiveness of the stock control, but what is held there is probably not ideal! Instead of considering the detailed ordering history for each item (and there are some interesting causes for stockholdings) the inventory can be classified according to why it is now in stock. An examination of the factors which give rise to the stock in a typical store reveals causes that are detailed in the following sections.

Purchase order quantity

There are several aspects which affect the ideal batch size. For example, most stores have items which are bought in bulk but sold in smaller quantities. At any time there is a balance of receipts in stock – a half empty box or pallet. This is usually an acceptable situation for keeping stocks at the right level, but that stock nevertheless still contributes to inventory value.

The size of logistic inventory is often determined by the transport methods chosen and the batch size of delivery which results. Changing the delivery method can change the stock levels in the stores.

Where items are being manufactured, the rate of production in one stage of the process may be greater than at other stages. The ideal solution to this would be Kanban (see section on *Pull systems* in chapter 4), but often there is a stock of items manufactured slowly awaiting to fill a process load. Alternatively a bulk process, or one where items are produced at a fast rate, will create stocks which are subsequently used up at a much slower rate.

Safety stock

In many situations customers do not provide information about their demands far enough in advance. To compensate for this problem two tactics are available. The first is to organise the customer to give more forewarning. Is this impossible? Manufacturers of specialist, make to order or highly sought after products may make their customers wait until the product is ready. In more competitive situations, however, the customer has more power and may not be prepared to wait. This puts pressure on suppliers to reduce the lead times and to forecast demand better. With better knowledge of the customers' demand, the unexpected peaks would no longer be a surprise and stocks could be varied to cover these occasions. The second option is to hold sufficient stock to cope with unexpected or excess demand. Safety stock (or buffer stock as it has been called) is there to cover our inability to predict demand. The inventory manager has an investment choice. Either invest in inventory or invest in information. Those who can accurately predict the customer requirements do not need safety stock.

A secondary role for safety stock is to compensate for failure on the supply side, non-delivery of purchased goods, information failures, technical breakdown or even industrial action. You cannot be too careful! Generally the approach to these situations is to monitor the situation continually and there are several standard ways of doing this (see lead times and schedules in chapter 7).

Market change

Customers change their mind, contracts are lost, markets vary. These all cause excess or even obsolete stocks, but this should not be a significant value. The changes that cause a buildup of inventory are either step changes or gradual changes. A gradual change is the most common and leads to many minor excesses of stock. The cause is normally a forecasting method that is not reactive enough, and the simple cure is often to use a more advanced forecasting method. Methods discussed in Chapter 11 and used in the right way can avoid this situation. A sudden reduction in demand can only be anticipated through understanding or information. Input from sales personnel is important in avoiding excesses as well as stockouts. Sales can warn of impending major changes, even if they are not good at identifying the size or the timing of changes. This is the best way to avoid having items remaining after demand is satisfied.

Obsolescence

In addition to excess stock caused by customers changing demands, a significant amount of obsolete stock can be caused within the company itself. Consider several situations:

1 The marketing department put an item on promotion without informing the inventory controller. The demand increases as the promotion starts, inventory control see demand escalate and stocks run out and therefore order some more. The promotion ends just as the new stock arrives. The company now has three years' worth of stock cover at the normal usage rate and major customers have already bought extra during the promotion.
2 The design department has replaced the old version of a product by an improved one, the sales team were extremely pleased with the new product and started selling it immediately. However, a large amount of the old product left in stock was not sold.
3 The marketing department has just launched the new season's catalogue. It contains some new items and some old ones have been deleted. There are still good stocks of some of the deleted items.
4 The sales director has bought a job lot of items which are going to be sold as a new line.

All these are common occurrences in some businesses. They all cause self-inflicted obsolete stocks and can be avoided by improving communication between departments. The responsibility for organising this communication revolution lies with stock control, since the other departments rarely understand the challenges of inventory management or the concept of lead time.

Poorly defined responsibility

When buyers procure to optimise purchasing costs and operations and sales sell to suit customers, in the middle are the stores, accumulating stock and controlling the conflict between the two activities. Stock control in this case becomes simply recording and storing. Inventory management should have the responsibility for assessing the demand pattern, determining the appropriate stock levels to support this and then instructing purchasing what and when to reorder.

Planned inventories

Cyclic demand, on a monthly, quarterly or an annual basis may be supported most economically by constantly holding high stock levels. In the stores there may be stock which is building up to satisfy a demand event or which is stored to cover a definite requirement. The feature of these is that they cover situations where the demand events are certain and in excess of the supply capabilities, as for example in industries where there is an annual holiday shutdown. Stock has to be provided to cover customer demand over this period and it is not possible to acquire the extra quantity at the last minute to cover the shutdown. The stock is therefore gradually increased over several weeks to maintain continuity. This effect can be complicated if different stages in a supply chain take holidays at different times.

Layout and location of stores

The effect of a stock location must be analysed in the context of the whole business. In a company where the store is situated inconveniently or a long way from the users, the users tend to keep their own stock. In fact they might keep some just in case the stores have run out when they want some. This lack of cooperation and control can result in satellite stores being set up and excess stock being held.

A worse situation results if different departments order the same items independently. In this case, there are totally independent stock systems, usually controlled by amateur stock controllers whose major interest is another aspect of the business and, as a consequence, stocks are excessive. Fortunately where this happens the items are normally low value consumables.

Where a company has more than one store (such as in distributed warehousing), a similar range of stock items are held in each, resulting in duplication of stock. If there are slow moving items in one store (try and find a store without slow moving items!) then the same item can be found in the other store, still moving slowly and total stock across all the stores can exceed the annual demand. This is a natural result of distributed stock. The safety stock required will increase as the square root of the number of distributed stores, so that it would be better if managers thought carefully about setting up local stock points. The increase in stock (by the square root) is based on the assumption that the local stores are part of an integrated, well-controlled inventory system. If this is not the case in practice, then the total stock can go through the roof (literally).

Company strategy

Some stock controllers do not have complete authority to maximise availability and minimise costs. This may be because top management has imposed some constraints, such as an extra stockholding, through lack of confidence in the ability of the stock system to provide adequate customer service. It may result from a commitment to customers to keep an agreed quantity locally or to keep consignment stocks at customers' premises. In the normal course of events the stock control system will provide the desired availability by adjusting the control parameters and should not need to be overridden by management edict. It is up to inventory management staff to instil confidence in their ability to control stock excesses.

System and control

The time taken to process information can substantially affect the level of stockholding. Consider a stores where an issue is being made. This causes the stock to fall below the ordering trigger level. How long does it take before the supplier is informed that a new delivery is required? On a conventional order processing system the procedure is:

- Enter issue on system (delay – 1 hour?).
- Print out list of items that need to be ordered (next morning?).
- Confirm that items do need to be ordered (rest of the day?).
- Print purchase orders (next day?).
- Obtain authorisation for purchase (when director is next available!).
- Send order to supplier (2 days by second class mail?).

These processes could easily take a week or much longer if the individual lines are batched up into a large order. If the supplier is a local stockist, the item could arrive on the same day, but the total supply lead time would be 1 week (internal) + 1 day (external), and the stock level would have to be correspondingly large to compensate for this lead time.

Slow systems cause extra stock

The solution is obviously to use the computer systems properly, without paperwork and to use communications technology, (fax or e-mail) for ordering.

If the stock is analysed according to the sections detailed above, the causes of stock excess become obvious. Analysis will show that actual stock is not held in the right balance to meet marketing and financial targets in the most effective way. By establishing a reason for this, a plan can be devised and action taken to ensure the lowest stockholding levels necessary to meet company targets.

Initially, the reason for stockholding has to be determined. It could be to achieve good customer service, cheap purchase, bulk discount, security or a large number of other reasons. The stockholding strategy depends directly on the company objectives and policy. Company policy defines the service which it intends to provide to customers, what investment to make in stock, how orders are placed on manufacturers or other principles on which to base supply plans. This is the first ground rule of stock control – to have a policy worked out by top management to guide the operation. The more professional the stock control operation, the closer the policy needs to be defined so that targets are set against which the activity can be run. This is a generic, broad brush approach, but is useful in determining where to put the most effort in stock management.

2

Customer service

Meeting customer requirements

The focus on customer service has gradually changed over the years and suppliers are now becoming really interested in customer service instead of just talking about it. Customer service is a complex subject of its own, but there are two main aspects, namely customer relations and availability of service or items.

Customer relations

Customer relations is about keeping the customer happy. It requires interpersonal skills to ensure that the customers have the correct level of expectation of supply and that they are happy with their purchase so that there is potential for repeat business and wider sales.

This aspect of customer service has gradually emerged because it is the differentiating factor between many companies. The product can be similar from a wide variety of vendors, the vendor which is most successful is the one where the quality of customer relations is best.

Customers may have a perception of suppliers which varies greatly from their actual performance. It does not matter whether a supplier is good or bad, what does matter is what the customer thinks about them. Customers can look for a match between their own style of company and the style of the supplier. If there is a match, then relations are likely to be good; if not, then the coordination requires more work. The major factors affecting customer relations are in sales and sales order processing and are outside the scope of inventory control. However, there has to be the same customer orientated approach in stock control which, like the Company, needs to maintain credibility.

The second aspect of customer service, availability, is a key target for stock control. It requires technical management.

Measuring availability

Availability

The entire reason for stockholding is to have items available, but despite the key nature of customer service to the success of the business, some companies do not quantify proper stockholding, only taking notice of complaints or other oblique assessments of service. To manage professionally there is a need for solid facts.

For each item in stock the risk of stockout can be reduced by increasing the stockholding. The larger the investment in inventory for an item the better the service. This is shown in Fig. 2.1 for an individual stock item. The curve shows that it is relatively inexpensive to give a reasonable level of availability, but the value of stock increases rapidly when we try to achieve very good service. 100% service is not possible because it needs an infinite amount of stock! However 99.9999% availability is possible, but expensive and no-one can tell the difference from 100%.

The shape of the curve shows that there is obviously a trade off between investment and availability of items. For each item, the stock policy can be adjusted to give the appropriate service. The overall service to the customers, or other stores users, can be measured from success in meeting demands (on time availability) on a daily, weekly or monthly basis. It can be calculated by adding the service provided on all the individual items taking into account the issue frequencies for the total range of items. We can define availability as:

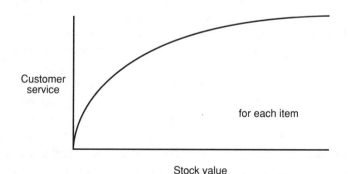

Figure 2.1 Availability of a stock item.

$$\text{Availability} = \frac{\text{Demand satisfied}}{\text{Total demand}}$$

This is the criterion which should be applied to all areas of inventory to monitor how well the investment in inventory is doing.

When the whole range of stock items is considered, the minimum stock levels result from all the items being under control at their optimum service level. Basically, each item should be on the same curve in Fig. 2.1. The only way to reduce the stock beyond this balance, while maintaining service, is to improve the total control method. After all, the aim is towards the top left corner of the graph – high service with low stock across the whole of the inventory.

Measurement of availability in practice

Within the definition of availability outlined above there are a number of ways by which the service level can be measured, depending upon the type of demand.

For companies supplying single items to individual users as a result of individual telephone or mail orders, then the definition could be:

$$\text{Availability} = \frac{\text{Total number of items supplied}}{\text{Total number of items ordered}}$$

For a department which is sending large export orders, these are often covered by a 'letter of credit' or the whole shipment must be sent in one crate. In this case the service level would be:

$$\text{Availability} = \frac{\text{Total number of complete orders supplied}}{\text{Total number of orders}}$$

For businesses where the demand does not have to be satisfied immediately, care must be taken to measure availability at the time required for despatch, not at the time of receiving the order. This gives the inventory controller the opportunity to fill the demand from elsewhere rather than from stock.

Care must also be taken concerning the definition of 'on time' delivery. Delivery should really be measured as the time the customer receives the stock rather than when it leaves the stores. 'On time' is when the customer needs the stock. Where there is a history of poor delivery the customers are likely to request a delivery date well in advance of their real need, in order to have a chance of receiving the items when they need them.

The type of customer also affects the measurement of availability. For

Table 2.1 Pitfalls of measuring availability

Part no.	Demand	Shortage	Service level (%)	No. of orders	Shortage	Service level (%)
A12	1000	10	99	5	1	80
B25	10	5	50	1	1	0
Average			74.5			40
Sum	1010	15	98.5	6	2	66.7

example, if a customer requires 10 and the stock available is 8, what customer service can be provided? Is the answer:

- 0% because the demand was not completed (non-stock/manufacturing answer).
- 80% because 8 out of 10 can be supplied (normal answer).
- 100% because the balance can be provided before the customer needs them (for stockholding customers).

Different situations have different priorities. To have an appropriate measure of availability, the service must be defined in one of the above ways, the one which suits most customers. In Table 2.1 only two items are being considered. The demand for item A12 was 5 orders totalling 1000 items of which 990 were sent on time and 10 items were not available for one order. This means that 80% of the orders (and 99% of the items) were sent on time. The second item is a slow mover and there was only one order (for 10 items) and only 5 were despatched on time giving a 50% service of parts (and complete failure in terms of complete orders). How good is the service? Out of 1010 items requested only 15 were short, so the number of items despatched was 98.5% of the total. However in terms of item order fill, deliveries of item A12 were 99% full but B25 was only 50% full, giving an average per line of 74.5%. If the customer is looking for complete orders, then out of the total of 6 orders two were short, corresponding to a 66.7% success rate. As an order fill per line, this was 80% for A12 and 0% for B25, so an average was 40%. This illustrates the different assessment of levels of service which can often occur between customers and suppliers. In this example the supplier is claiming 98.5% delivery on time, while the customer is measuring it at 40%! It is important to agree a common goal between suppliers and customers and to establish what the customers need in terms of availability. The appropriate measure can then be used for monitoring.

Availability policy

Customer service can be structured or focused on a particular group of customers or market sector using alternative availability policies. Options which could be used are:

- The same availability across all the products,
- Minimising the total cost of service,
- Concentrating on the most valuable customers,
- Enhancing service on the most sensitive products,
- Greatest availability of the most profitable products, or
- Better service on the major turnover items, reduced service on slow movers.

These alternatives will provide different availabilities for different products, with a management decision setting availability levels. Whichever alternative is chosen, proper control of stock requires the inventory manager to monitor availability regularly.

Back order measurement

The measurement of availability discussed in the section on *Measurement of availability in practice* should always be used, but a further measurement is necessary for sophisticated stock control. Once an item has missed its delivery date, it is out of the calculation. This introduces a further factor into the equation: the priority given to satisfying back orders. Consider a case where there are two demands for an item, one which is past the due date, and one which is to be delivered on time. If such a case arises, the availability target calculation requires us to fulfil the current order, not the old one, since the current one improves the availability factor, but the other has already failed. In practice there is normally a need to give precedence to customers whose delivery is late. Therefore a more rigorous measure of availability would be to measure how many orders are late and how late they are.

An example of how this can be approached is developed from Table 2.2 which shows the despatch position for two companies, A and B. Which company is providing the better service? From our service level (level of fill) approach company B is the better because it gives 30% level of fill (LOF) against 10% LOF for company A. However there is a better chance of getting the items quickly from company A. Perhaps A is better after all.

A measure of customer service which takes into account the overall performance, not just the 'ex-stock' part, can be developed by using a weighted

Table 2.2 Comparison of despatches for companies A and B items despatched in week 37: analysis of lateness of orders despatched

	Company A	Company B
Delivered on time	10	30
1 Week late	60	15
1 Month late	20	40
3 Months late	10	15
Total despatches	100	100

Table 2.3 Distribution of arrears

Time overdue T (weeks)	Number of items N (items)	Weighted average $T \times N$
1	8	8
2	6	12
3	4	12
4	2	8
5	1	5
Total	21	45

average. In Table 2.3 the arrears, (or back orders), have been analysed by categories of lateness. By multiplying the number of orders outstanding by the weeks of lateness a back order week figure has been calculated. This gives an overall figure of merit of 45 arrear weeks, corresponding to an average for arrears of 2.1 weeks per late item. By monitoring weeks of lateness against a target level, inventory controllers can assess the level of service achieved. Obviously, the best practice is to ensure that the bulk of demands are satisfied on time because those few items which are long overdue give a negative contribution to the figure of merit. If the back orders are satisfied rapidly, the level of service can be measured in units of days or hours rather than weeks.

In practice the order backlog viewed on the system may not be all true demand. It often consists of real arrears which are customer orders slightly overdue and a real priority and virtual arrears which are often long standing customer orders without any urgency on them. These virtual arrears are orders which the customer does not really want, or where they have bought an alternative instead and not cancelled the order for the original item. Outstanding orders need regular review to ensure that virtual orders are either rescheduled or removed. If virtual arrears build up they have a distorting effect on the back order measurement.

The extent of arrears can be measured in several ways. The number of items in arrears is the simplest measure and the aim is to focus the control on the important items. Measurement in terms of value is not usually recommended since the aim is to provide customer service. Routine assessment of arrears should be carried out in a similar manner to the monitoring of availability.

Demand management

Order processing

Stock should, of course, fill customer demand. In practice this includes both having stock available and despatching it. The physical handling of the stock is the responsibility of the stores operation, while in many companies the actual order processing forms part of the stock control function. Demand management consists of a series of steps:

1 Order receipt
2 Order processing
3 Estimating delivery times to customers
4 Production of despatch information
5 Customer feedback.

Order receipt

This is the initial compilation of orders into a form which can be used to despatch the goods at the right time to the customer. This information must be (i) complete, (ii) specific and (iii) accurate. The information can arrive by post, telephone, electronically or by word of mouth. A first requirement is that the company has a good item identification system which enables sufficiently unique identification to satisfy a discerning customer. This has to be formalised into data for processing.

Stock control requires an identification code for each item, ideally provided by the customer. To assist this, some companies provide customers with catalogues and order forms to enable customers to specify precisely. It used to be customary for counter sales customers of all types not to know code numbers and to rely on stores personnel to remember them. This situation is no longer acceptable – if customers want the correct item, then they should expect to specify their requirement properly. An order form not only enables the correct data to be collected, it also presents it in a standard format, which can be arranged to match computer input or stores picking procedures.

Order processing

This is the entry of the order into the system. A second requirement is that the processing of orders translates the customer order into company information reliably, speedily and without creating errors. This process has to be fast, simple, and above all, accurate. For computer systems this should be done as a one-stage process with the source (phone, fax, letter, etc.) being input directly into the system. This avoids errors and delays. Where printed customer orders are required, these should be printed out from the data input to the system.

For non-computer applications the source document (or a copy of it) should be used for stock picking. With technology, EDI (electronic data interchange) can create a demand in the system without any work. This is good as long as the demand is easy to fulfil. Stock controllers need the opportunity to vet orders, which means that all EDI orders should be formally accepted for despatch. Application of the 'manage by exception' principle suggests that most orders are easy to satisfy and can be accepted without question, automatically. It is only those orders for fast delivery, or for large quantities, or for non-standard items, which need control. Criteria can be set up to identify these, so that the majority, the other standard orders, can be accepted without question. The stock control parameters for establishing these criteria are discussed later.

Estimating delivery times

This is the activity which causes a high level of customer confidence to be gained or lost. Promises have to reflect the real stock and replenishment situation or else customer service will be poor (see section on *Estimating delivery times* p 26).

Despatch information

This will consist of two stages:

- picking and delivery note production and
- customer invoicing.

These stages should be carried out sequentially: the financial (invoicing) systems should not be allowed to interfere with the stock movement

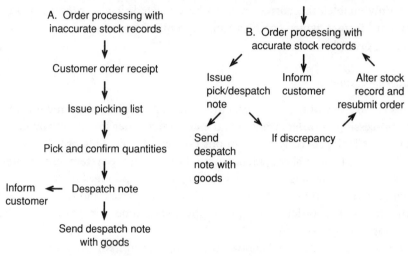

Figure 2.2 Stock item order processing.

(despatch) system, as invoicing is not part of the stock control system. Two alternative paperwork processes can be used. A delivery note can be prepared as soon as the order is processed, as long as the stock records are accurate as shown in Fig. 2.2B. Where records are suspect, a safer procedure is the one shown in Fig. 2.2B. This usually causes extra paperwork.

Customer feedback

To report on progress and to keep the customer feeling involved in the supply is very important from a customer relations standpoint (see section on *Meeting customer requirements*). Customers often require acknowledgement of order and confirmation of price as well as a despatch note and invoice. This confirmation should form part of the normal information processing procedures.

Consuming forecast demand

In almost all situations business works to a forecast until the real orders arrive. However, real orders rarely match the forecast and the question is raised as to whether an actual order received is really part of the forecast.

The ideal situation is when a company accurately forecasts demand and makes that quantity available. As customer orders arrive, available quantity

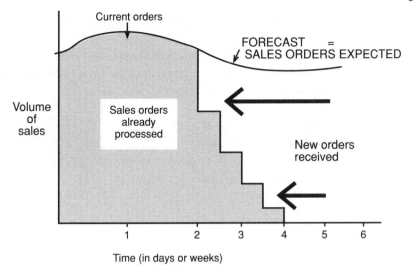

Figure 2.3 Consuming the forecast.

is allocated and despatched. When the quantity for the current period has all been allocated, customers are provided with items in the next period. This situation is shown in Fig. 2.3. Here the forecast in periods 1 and 2 (shown as the curve) is fully allocated and there are orders for periods 3 and 4 which have already been received. The demand in periods 5 and 6 is all forecast as no orders have been received which need satisfying so far ahead. Orders are constantly being received and product is allocated to these customers. This consumes the forecast quantity and builds up the allocations. This approach ensures that a company does not allocate more than has been provisioned through the forecast. In some fast turnaround businesses with period length of a day or less, this situation is acceptable and almost inevitable. In other businesses more flexibility is required and there has to be

• Safety stock to meet high demand or
• Extra supply sources available to meet peaks or
• Resources reserved for emergency supply (e.g. allocation of only 90% of capacity in manufacture leaving 10% for last minute orders).

Where a company agrees to fill every order from customers at a time to suit the customer, it is important to forecast carefully, otherwise delivery achievement will be poor, much management time will be spent chasing problems, panic purchases will be made and extra deliveries will be required, causing costs to be high.

The differences between forecast and actual demand are measures of the quality of the forecast (see chapter 6). Actions arising from order processing, besides accepting the customers' request, could be:

- To arrange increase or decrease in stock levels.
- To bring forward or delay supply.
- To offer an earlier or later delivery date to customer.
- To reject the order, or offer alternative item.
- To provide part shipment.

An example

A supplier of fasteners has trouble with widely fluctuating demand levels on some products. This is because the customers think 'I need a box of fixings for this job' and order enough to do about two months' work. Consequently the supplier must provide for 2 months' demand and then nothing. This is acceptable if there are many customers, but demand can be very erratic for the slower moving fasteners. By understanding the supply problem, the customer can be encouraged to order enough to start the job off with sufficient supplies, while the supplier, who understands the user's situation, can provide the range of fasteners but is able to hold much less stock. Of course, the situation must be carefully monitored and the balance must arrive before the initial stock has been used up. Large quantity orders can be filtered out using simple criteria (see chapter 6), so that real demand priorities can be fulfilled. Demand can thus be managed through the order processing activity.

Estimating delivery times

For supply to order items, and for stockouts, an estimate of the delivery date is usually required. Sales are often not aware of the status of the supply and the order book, and so are badly placed to give more than a blanket estimate of delivery. Stock control or purchasing do have the information and can provide dates which are based on the actual current situation.

It is the responsibility of inventory management to provide delivery and availability information in support of customers' orders. The time taken to provide an item for a customer depends on the balance between stock currently held and on order and the quantity already allocated for customer requirements (see chapter 12). If the order being accepted is allowed to use items already allocated to other customers, then the order can be fulfilled sooner. In general the priority sequence is 'first come first served', but there are often cases where certain customers gain priority. Sales may consider that promising items independently of the ability to deliver will secure orders and keep inventory control people on their toes. This can lead to a mistrust of the optimistic delivery promises made and poorer customer service. The solution to this situation is for inventory control to provide customer delivery dates using priority rules specified in advance by marketing. Where there are exceptional circumstances, priorities can be altered within the total availability profile. The request to 'ring the supplier and get a faster delivery this once' should be dismissed on all but the odd occasion, since it usually results from sales people not being able to negotiate successfully.

Monitoring delivery dates can become complex. Dates which can be considered appropriate are:

1   When the customer placed the order
2   When the customer requested delivery
3   When the customer needed the item (usage date)
4   When delivery was originally promised
5   When delivery is now promised
6   When the item is despatched
7   When the item is received by the customer

and it is also useful to know

8   Who changed the delivery date?
9   How many times has it been changed?
10   When the delivery date was changed?

The importance of these dates varies from one market to another and one company to another. There are so many dates and potential monitors that companies have to choose which ones are important in their own circumstances. Providing too many management controls is confusing and dilutes the important ones. The prime controls would be delivery on time measured as '7–2' and '7–4'.

The difference between the customers' request for delivery and expected delivery dates (2–1) is their perception of lead time. This may be a general indication of market expectations. The difference between the customer order date and the date they received the goods (7–1) is the actual customer lead time. Unfortunately, most companies measure their despatch dates (6) rather than the customer receipt dates, so there could be a difference between the two views of delivery performance. Many carriers now have sophisticated tracking systems which have the potential to provide suppliers with customer receipt information to use in their control systems.

Customers may leave an interval between their requested supply date (2) and the time they actually need the items (3), to compensate for delivery problems. If information is available to monitor this, it can be used to improve collaboration between supplier and customer. The monitoring of delivery performance, as with all other inventory monitoring systems, exists to provide the best factual information for management and control. Monitoring systems should therefore differentiate between when the customer wanted the goods (2), when the delivery was originally promised (4), and when it is now promised (5). It is useful to measure credibility (7–4), to ensure that the supply systems are working correctly. Measuring performance against current delivery promise (6–5) can fool a company into thinking falsely that it is performing well. The current delivery promise may bear no relation to when the customer requires the item, and may be the latest in a succession of promises as the delivery date slips back. The estimated delivery date can often be changed so that it seems that 100% performance is achieved. Therefore monitoring success against current delivery promise is not likely to be a reliable measure of effectiveness. Monitoring of slippage – the number of times the delivery promise has changed (8) can also be revealing in showing whether the company is providing customers with good information and service. However, the reason for late delivery against the original plan could be simply that the customer does not require the goods until later. It is therefore useful to examine (9) what proportion of the changes are due to poor supply and how many are due to customers changing their mind.

3

Managing the inventory

Using Pareto analysis for control

Applying effective control

This simple principle is embodied in Pareto's law, which is illustrated by the curve in Fig. 3.1. It is also called the 80/20 rule because 80% of the effect is provided by 20% of the cause. In this illustration 80% of stock value is caused by 20% of stock lines. The principle can be applied to many different areas of activity: 80% of the purchased items come from 20% of the suppliers, and 20% of sales lines give 80% of turnover. For a warehouse, 80% of the space is occupied by 20% of the lines. The simple fact makes it obvious which lines should be received little and often.

Pareto analysis is the technique which forms the basis of inventory control thinking and is an important management principle which can be applied to minimise effort and to obtain best results. It can also be applied to time management, credit control and many other areas of control. To gain best control, effort has to be directed to the most important areas. The Pareto curve (see Fig. 3.1) is often called the '80/20 rule', but the values can be read off at any convenient point. For example, the graph shows that 50% of the product lines account for 97% of the sales (or that the other 50% only provide 3% of the sales – a worrying thought). The shape of the Pareto curve arises from the range of volumes and values combined in a statistical distribution. The shape of the curve does not always give an 80/20 relationship exactly, but this does not affect the principle of applying Pareto analysis to inventory management.

Stores carry a wide variety of items, with a stock record for each. Some have high value and others are very cheap. The high value items are normally controlled tightly, whereas the low value items are not treated as carefully and are issued in bulk in approximate quantities. Most effort should

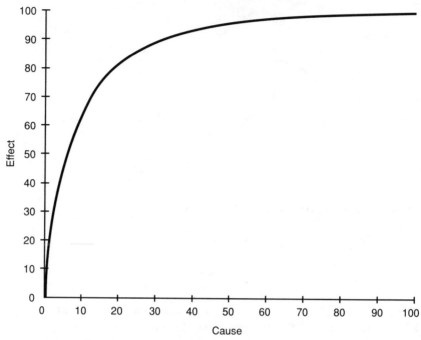

Figure 3.1 Pareto curve.

be put into managing items which are most important for achieving inventory targets. In inventory control the best results are gained by organising effort correctly. There is insufficient management time to maintain detailed control of all individual items. If the immediate aim is to reduce stockholding costs, then studying the stock of low value items is unlikely to be the best place to start unless the sales volume is large. If service is the aim, then attention to a few fast moving lines often provides the bulk of the improvement required.

Stores contain items ranging from main products to washers and labels, with a stock record for each. High stock value items need to be closely controlled, whereas minor items need not be treated as carefully. To control the resources of the company most effectively, effort and controls should be biased towards high cost areas. Pareto analysis formalises efforts to do this. It states that the majority of the effect is produced by a small proportion of the cause (80% of the effect is due to 20% of the cause). The application of this to a stores stock control means that 80% of the total stock value is made up from 20% of the total stock items as stated. The other 80% of stock items contribute only 20% to the total inventory value (shown in Fig. 3.1). In a stock reduction exercise the majority of cost saving will be gained by decreasing stocks of those few major items.

Example

Consider a stock of 12 000 types of items in store. Pareto's law shows that for a stock value of $800 000 we find that 2400 items account for $640 000 of inventory. The remaining 9600 items are worth only $160 000. Therefore by concentrating on the 2400, control over the total value will be tight. If 2400 is too many to review individually then the Pareto curve (Fig. 3.1) shows that 5% of items account for 55% of cost so 600 items contribute $442 000 to the total stock costs. Again by working on these 600 items carefully the overall stock value can be controlled or decreased.

ABC analysis

Pareto analysis by the current stock level is good for reducing stock levels, but a more consistent classification is required when focusing on the management of inventory. The current stock does not necessarily show which items are important for the business. In fact there may be some important items where the current level of stock is low because stores are awaiting an impending delivery. On the other hand some items may have a high stock value simply because no-one is buying them. It is therefore usual to rank the items according to the annual turnover. The annual turnover is given by

Annual usage × Unit cost

It is not too important whether the unit cost is the standard cost, latest cost or an average as long as it is consistent across all the items. Annual usage has to be adjusted in the case of new or obsolescent items to reflect the future expected demand rate rather than the historical one.

Pareto analysis of this data shows that 80% of the value of demand is for 20% of the moving items. (There are often a number of items in stock for which the demand is zero, and therefore not included in this turnover analysis.) For some businesses the 80/20 rule is not obeyed exactly, but the use of Pareto analysis is important for all inventory.

To use Pareto analysis properly requires classification of stock by issue value and the simplest way is to use ABC classes. These can be defined as:

- A = 10% of stock numbers, giving 65% of turnover
- B = 20% of stock numbers, giving 25% of turnover
- C = 70% of stock numbers, giving 10% of turnover.

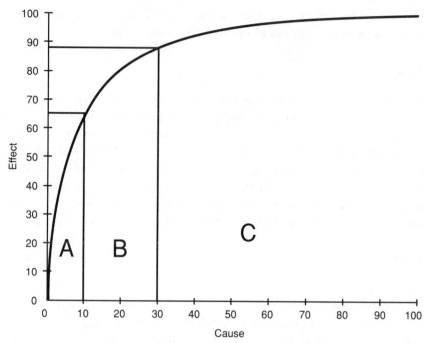

Figure 3.2 ABC analysis.

This is illustrated in Fig. 3.2.

It is important to ensure that ABC analysis is based on turnover, but less important that the exact percentages are adhered to. In some instances a further classification (D) is useful to include a large number of very low turnover items. This enables the number of stock lines included in the A, B and C classes to be reduced to manageable numbers.

The purpose of ABC analysis is not to provide different types of service but to provide service with the least amount of cost and effort. Different systems of control are used for the three categories of stock. As the A items carry the most value, accurate systems are required to control them. On the other hand, the C items are low turnover value but form the bulk of the inventory. For these the main requirement is to ensure that stock is available to meet demand.

The control requirements for each category are shown in Table 3.1. Category A items have a disproportionate amount of time and effort used on them and have to be controlled tightly using systems in conjunction with market expertise and product knowledge to maintain stock at the lowest

Table 3.1 ABC inventory control

Characteristics	Policy	Methods
A Items		
Few items	Tight control	Frequent monitoring
Most of turnover	Personal supervision	Accurate Records
	Communication	Sophisticated forecasting
	JIT approach – balanced safety stock	Service level policy
B Items		
Important items	Lean stock policy	Rely on sophisticated system
Significant turnover	Use classic stock control	Calculated safety stocks
	Fast appraisal methods	Limit order value
	Manage by exception	Computerised
		Management & exception reporting
C Items		
Many items	Minimum supervision	Simple system
Low turnover value	Supply to order where possible	Avoid stockouts and excess
(Few movements or low value items)	'Large' orders	Infrequent ordering
	Zero or high safety stock policy	Automatic system

appropriate level. For category B items computerised techniques are most appropriate. The number of items involved and the lower values make it a waste of time to use specialist skill which could be working on category A. The computer can maintain control through statistics and deal with the complex calculations using the computer forecasting models which are most important for B items. The use of management by exception is also important for the B class. The minor sales items, category C, should be controlled by a simple system which enables supply to be obtained with a minimum of administration. However the control system for C items must be reliable and not result in stockouts or large excesses. An investment in extra stock of C class items is inexpensive but can greatly simplify the problems of controlling large numbers of stock lines. This is an appropriate policy for the faster moving C class items. For the very slow moving, higher value C class items a purchase to order policy should be adopted where possible, or if there is only one customer, they can hold the stock themselves and be responsible for reordering as required.

The most effective stock control systems are based upon ABC analysis combined in a common sense manner with the other techniques still to be discussed here. ABC analysis is the basis for the total control of stock. It is also used as the basis for perpetual inventory stores control where annual

Table 3.2 Example of Pareto analysis

Item	Annual usage (units)	Unit cost ($)	Annual turnover ($)	Annual turnover (%)	Rank
A12	21	7	147	2.1	5
B23	105	11	1155	16.2	2
C34	2	15	30	0.4	10
D45	50	5	250	3.5	4
E56	9	14	126	1.8	6
F67	397	12	4764	66.8	1
G78	5	8	40	0.6	9
H89	500	1	500	7.0	3
I90	11	4	44	0.6	8
J01	3	25	75	1.1	7
Total			7131	100	

Table 3.3 Classification by usage value

Item	Annual usage (units)	Unit cost ($)	Annual turnover ($)	Annual turnover (%)	Rank	Class	Cumulative percentage
F67	397	12	4764	66.8	1	A	66.8
B23	105	11	1155	16.2	2	B	83.0
H89	500	1	500	7.0	3	B	90.0
D45	50	5	250	3.5	4	C	93.5
A12	21	7	147	2.1	5	C	95.6
E56	9	14	126	1.8	6	C	97.3
J01	3	25	75	1.1	7	C	98.4
I90	11	4	44	0.6	8	C	99.0
G78	5	8	40	0.6	9	C	99.6
C34	2	15	30	0.4	10	C	100.0
Total			7131	100			

stocktaking is avoided by routine counting of a few stock parts each week. An example of Pareto analysis in action is given in Tables 3.2–3.4 and Fig. 3.3. Table 3.2 shows a number of different stock items, their unit cost and annual usage in terms of quantity of value. Items are then ranked in order of size of annual turnover value. This is displayed in Table 3.3 with the items in descending order. The cumulative annual usage value and percentage of turnover are also calculated. Items are then classified into A, B and C by assessing the number of items in each category and their percentage of total turnover value. In Table 3.4, classes A, B and C are compared as percentages of number of items and total value. The table shows

Table 3.4 Summary of ABC analysis

Classification	Percentage of items[a]	Percentage of value	Value per class
A	10.0	66.8	4764
B	20.0	23.2	1655
C	70.0	10.0	712
Total	100.0	100	7131

[a] ABC analysis is carried out only for items with usage.

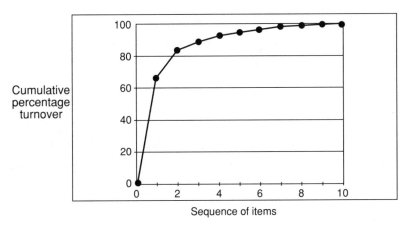

Figure 3.3 The cumulative Pareto curve.

that the seven items in Class C have a turnover of £712, whereas the only item in Class A has a turnover of £4764.

Saving time

Applying the Pareto principle is a way of balancing inventory, stock availability and critical resource spent on each item. How the law can be applied depends on what the critical resource is considered to be. The critical resource for all inventory controllers is time, because of the large amount of information required and the wide variety and quantity of stock held. The Pareto principle shows that 80% of the time is spent doing 20% of the jobs and a significant time saving can be made if a small reduction can be achieved in these jobs. They may be very frequent short jobs (such as keying stores issues into a computer) or more infrequent, long-winded jobs (such as writing a major management report).

Practical considerations in using ABC analysis

The ABC classification is a simple tool to enable the stock manager to control a large number of items in a limited amount of time. This simple approach is one of the most powerful tools employed to reduce stock value and to decrease the workload of busy inventory managers and purchasers. Experience with this technique over a large number of companies has suggested that there are some pitfalls and some practical ways to circumvent them. The situations which arise are typically:

1 Too many A class items.
2 Large numbers of lines (D class).
3 Non-moving lines (O class).
4 Fixed stock level items (F class).
5 Non-stock items.

Number of A class items

A class items are supposed to be reviewed on a daily basis or a weekly basis at least. To be practical, there has to be an upper limit of, say, 300 A class items per inventory planner. Even where the planner manages many thousands of items, the number of A class has to be kept small.

D class items

For normal inventories of 3000 lines, the ABC classification works well. Where there are over 10 000 part numbers there should be a modification to the classification system so that the vast bulk of low value turnover items are dumped into a further class, D. The class contain the lowest turnover lines, say 50% of the active item numbers which contribute only 2–3% of the total turnover. The ABCD classification now is as follows:

- A class: 5% of moving lines (300 per planner maximum).
- B class: 10% of moving lines.
- C class: 35% of moving lines.
- D class: 50% of moving lines.

The classification can alternatively be carried out by turnover value, for example:

- A class: 45% of turnover.
- B class: 30% of turnover.
- C class: 22% of turnover.
- D class: 3% of turnover.

These classifications can be varied to suit the exact shape of the Pareto distribution for the inventory to be managed. The principles of the technique do not change even if the distribution is not 80/20.

Non-movers

So far in the discussion the moving inventory has been classified and the non-movers have been ignored. Companies do sometimes have a need to keep non-moving items or items where the movement is so slow that they appear as non-movers in the recording systems. These items will probably be subject to the stock cleansing discussed in the section on *Turnover of stock*. In the interim they should be provided with a separate classification. Normally they are given class 'O' or 'X' which signifies that they should not be ordered again.

Fixed classification

As Pareto analysis is to be used for ordering, there are a few items where the stock level should not respond to usage rate. (Unfortunately in many older stock management systems, the stock control parameters are all like that!) For example the employment of two new maintenance fitters in a factory requires them to be kitted out with protective clothing, tools, tool boxes and a variety of items. The use of these items is likely to fall rather than rise as a result of that action since there is less likelihood of further recruitment. By putting these items in a separate classification (say F), they can be segregated and identified. The system can then identify that the stock level parameters from classification F are not updated.

Non-stock items

The decision about which items are in the stock range is an arbitrary one and depends on the particular inventory policy and market conditions. Customers do not usually consider their requirements as 'stock' or 'non-stock' items and the difference to them is only in the lead time provided by the supplier. In fact non-stock items are continuously being taken into the stock range and stock items are being deleted. For this reason it is useful to include non-stock items in the ABC analysis. This can either be done by including them within the ABC classes, or more commonly, by having a separate classification (Z?) for non-stock items. This separate classification then has the feature that no stock is ordered from suppliers unless the stock cover is negative (i.e. there is a customer requirement but no stock). It is useful to include all goods in the inventory management system as it provides unified records identification and control over all goods and makes management and analysis easy.

Stock cover

Turnover of stock

Current stock levels in the various stores throughout the company may not all be at ideal levels, as we have seen. The purpose of controlling the inventory is to drive the stocks towards their proper level which is determined by the characteristics of supply and demand patterns. The major factors are:

- Supply lead time
- Average demand rate
- Variability of demand
- Supply frequency
- Customer delivery time allowed

There are also practical considerations such as:

- Reliability of the supplier
- Criticality of the item
- Availability of item from other sources

The concept of 'balance' is most important in ensuring that the maximum service is produced from a minimum of stockholding cost.

The best level of service will be provided if there is an equal chance of all items being available for the customer. High stocks of one item and low stocks of another will reduce the overall availability and increase the inventory cost. The percentage availability target should therefore be the same for all items in the stock range.

If the cost of managing the inventory is also considered, the value of items being ordered and controlled becomes important. Valuable inventory management resources should be confined to the most cost effective jobs. This means that time should be spent on high turnover value items and not wasted on items whose value is insignificant. The best balance of inventory leads to an optimisation of costs and service over the full range of stock lines within the time available.

The inventory performance of each item can be monitored using a figure of merit for stock balance. This is the 'stock cover', which is defined as:

Table 3.5 Stock cover

Item code	Stock (units)	Annual usage (units)	Stock cover (weeks)
1P1	250	2000	6.5
1P2	700	1625	22.4
1P3	500	400	65
1P4	15	1000	0.78
1P5	20	25	41.6
1P6	40	250	8.32
1P7	500	200	130
1P8	8	400	1.04
1P9	6	40	7.8
1P10	65	20	169

$$\text{stock cover} = \frac{\text{current stock} \times 52}{\text{forecast annual usage}}$$

This gives the result in terms of 'weeks-on-hand'. The same answer could be calculated for a month's predicted usage and multiplying the stock by 4.2 instead of 52. This is more convenient sometimes, but should only be used when demand is consistent. It is also convenient to use historical average usage in this equation (see chapters 10 and 11). A sample of stock items is shown in Table 3.5. Which of these items requires attention first?

Stock cover gives an insight into the priority for action. It is not an infallible guide, but it does indicate where review is required. The first instinct would be to look at 1P4 and 1P8, where the stock cover is small. However, there may not be a problem with these since a delivery may be arriving. At the other end of the scale 1P3, 1P7 and 1P10 have over a year's worth of stock, so that means of reducing this level will have to be found. Stock cover shows whether the stock is 'in the right ballpark'. If it is not correct, attention to the items which are obviously well outside acceptable stock levels will keep the shape of the inventory reasonable. Stock cover is used by inventory controllers because it is easily understood in terms of usage rates and lead times. Stock cover is the time in which the stock will run out at average usage rate.

As well as being a crude analysis tool for each stock item, stock cover is also an important tool for measuring the total inventory. Overall stock cover is calculated from the total value of stock divided by the annual issue value and multiplied by 52 to give data in terms of weeks. In many distribution stores overall stock cover is for between 1 and 8 weeks. Financial managers are often more interested in the use of funds and therefore measure the effectiveness of inventory management using 'stock turnover' or 'stockturn'.

This is just the reciprocal of the stock cover, taken on a value basis for the complete stockholding.

$$\text{stock turnover} = \frac{\text{value of annual usage}}{\text{value of stock}}$$

This calculation gives the number of times the stock would be used up per year.

Example of stockturn calculation

Value of stock in the stores is $150 000. Issues for the last 12 months amount to $900 000. Stockturn is therefore 900 000 ÷ 150 000 = 6. This means that the stock value would be used up completely 6 times per year so that the stock cover for the total stock will be 2 months, or by the stock cover calculation as

$$\text{stock cover} = 150000 \times 52/900000 = 8.67 \text{ weeks}$$

Stockturn is based on historical data and is used for financial reporting. Stock cover is an inventory management tool for planning stockholding and can be based on known data and the forecast usage rate so that the stock will meet the expected demand for the item. When the stock level is being assessed for accounting purposes, the ratio uses the historical usage rate, which enables a conservative view to be taken of the stock level. Although this sometimes leads to a divergence of views on the necessary stockholding, the assessed future demand should always be used when controlling stock or placing orders.

Setting stock targets

For good stock balance the stock cover of all the items should have similar value. In practice differences in the variability of usage and the order cycles leads to a range of acceptable values for the items. Stock cover should not be used for working out reorder levels – there are proper accurate ways of doing this (see chapter 7). Stock cover ratios can be used to calculate the broad ranges of weeks' cover which would be needed for inventory items. For instance it is unlikely that more than 52 weeks' worth of stock is planned for any inventory item, therefore a figure of more than 52 could be the boundary between 'OK' and 'needs attention'.

As inventory control should be tightest for the A class items, these are the ones which can be controlled down to lower stock cover figures (leading to the paradoxical situation of holding lowest stocks of the best sellers!),

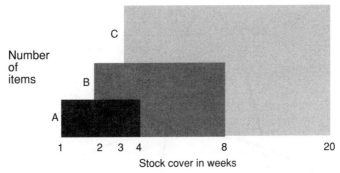

Figure 3.4 Ranges of stock cover.

whereas extra stock of the minor C class items adds little to stock value and significantly reduces the work of controlling.

In practice the stocks could have control limits to avoid extremes of inventory, and an allowable stock cover range can therefore be set by the ABC inventory classes in a ratio which is theoretically 1:3:7. An illustration of the acceptable ranges for stockholding of category A, B and C items is shown:

- A class   between 1 and 4 weeks
- B class   between 2 and 8 weeks
- C class   between 3 and 20 weeks.

A stockturn ratio (weeks of stock) for all the items in the category should lie in the ranges shown (see Fig. 3.4).

In theory the distribution of the population of items within a class is shown in Fig. 3.4. This figure also shows what the curve looks like in practice. (Note that the horizontal scale is logarithmic.) The theoretical curve allows for the variability of demand, usage of safety stock for some items and some slow movers, causing some of the items in a class to fall naturally outside the expected limits. The actual population curve when plotted against cover has an entirely different shape. Some items have high stock cover ratios, and since the stock value has to be limited, some of the other items have very low stocks to compensate. The shape of this curve is caused by the response systems initiated by low stock levels and the less effective action which normally is seen when the stock levels for some of the items rise over the maximum. The profiles of stockturn ratio in Fig. 3.5 are typical, even where inventory management practices are good.

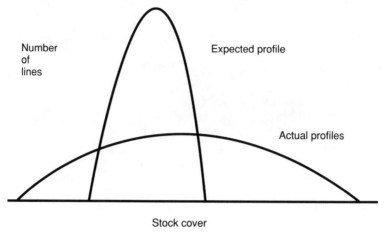

Figure 3.5 Expected and actual stock profiles.

The weeks of stock cover shown in Fig. 3.4 indicate how long it would take at best to reduce stock levels. If the stock reduction is to be made through C class items, the time taken to reduce the levels will be in excess of 20 weeks. For A class items the whole process should be completed within 4 weeks.

Pareto stock balance

Pareto analysis enables us to take the discussion of stock balance a step further. Each Pareto class, calculated according to annual turnover, can have its own target months of stock range as shown in the illustration of acceptable ranges of stockholding in the previous section on *Setting stock targets*. The use of these two techniques (ABC analysis and stock cover) together is fundamental in forming the basis of the stockholding policy. Another illustration of the power of the Pareto rule is shown when the overall value of stock has to be reduced to meet a new target. Say the inventory investment currently amounts to 6 weeks' of stock and this needs to be reduced by a minimum of 10%. ABC analysis of the inventory items would follow the pattern given in Fig. 3.5.

In Table 3.6 the A, B and C classes have average stockturns of 2, 5 and 11.5 weeks, and therefore constitute a well controlled inventory. If these amounts of stock cover appear to be over generous, then the substitution of the word 'day' for the word 'week' in the heading will suit all but those involved in JIT supply (see chapter 4). The classes contribute 65%, 25%

Table 3.6 ABC stock cover

Class	Moving lines (%)	Stock cover (weeks)	Turnover (%)	Weeks of value
A	10	2	65	1.3
B	20	5	25	1.3
C	70	11.5	10	1.2
Total	100		100	3.7

Table 3.7 Options for reducing stock value

Class	Moving lines (%)	Stock cover (weeks)	Turnover (%)	Weeks of value	Reduction in weeks of value	Number of lines involved
A	10	1.5	65	1.0	0.33	100
B	20	5	25	1.3		
C	70	8	10	0.8	0.35	700
Total	100		100	3.025		

and 10% to the inventory value, respectively. This means that the total stock values arising from the A, B and C classes are equal.

There are several options for decreasing the stock value. One way is to cut A items down to an average of 1.5 weeks of stock. This new situation is shown in the A class calculation in Table 3.7. The net result is a stock reduction of 0.33 weeks.

An alternative way to achieve the same result is to reduce the stock of C class parts down to 8 weeks (from 11.5 weeks). The new contribution to stock cover is 8×0.10 (=0.8) weeks of value, a reduction of 0.35 weeks.

Tighter control of A items is the better alternative because:

• The change required is less (0.5. weeks as opposed to 3.5).
• There are fewer lines involved, so less disruption of planning.
• Less work will be involved – 100 stock items instead of 700 (last column of Table 3.7).
• The effect will be more rapid.

This last point can be proved by looking at the stock cover for each of the ranges in Table 3.6. The average taken for stock cover for A, B and C, are 2, 5 and 11.5 weeks, respectively. If the classes were perfectly balanced and the supply was stopped, then A class would run out after 2 weeks, B class

Table 3.8 Number of purchase orders with ABC inventories

Class	Moving lines (%)	Order cover (weeks)	Turnover (%)	Average weeks of value	Orders per year[a]
A	10	1	65	0.33	5 200
B	20	4	25	0.50	2 600
C	70	10	10	0.50	3 640
Total	100		100	1.33	11 440

[a] per 1000 inventory lines.

after 5 weeks and C class after 11.5 weeks. Stock reduction through natural usage will take at least these respective times. As stocks are not often perfectly balanced, the true time to reduce the stock is normally over twice this long.

Purchase order patterns based on Pareto analysis have been devised to save stock value and management time. This is well illustrated by the ordering process shown in Table 3.8. Consider an inventory of 1000 different lines. If each line is purchased each month from the suppliers, then there are 1000 lines ordered each month, or 12 000 per year, and a stock cover of just over 2 weeks plus safety stock. (Stock goes up by 4.2 weeks' worth on receipt and then reduces to nil, so the average is 2.1 weeks.) Safety stock can be omitted from the discussion since it is taken to be independent of the order pattern. Now, if the A class items were ordered twice as frequently and the C class half as frequently, what would be the effect? The answer to this question is shown in Table 3.8.

The result is that:

- The total stock value is reduced from 2.1 weeks' cover down to 1.55 weeks cover (excluding safety stock in each case).
- The number of line orders placed is also reduced from 12 000 per year down to 9750.

This shows that a reduction in effort and a decrease in stock are brought about at the same time. These results are typical and are borne out in practice. The effect on customer service from these changes is negligible, because the safety stock remains the same in each instance. (The availability is based on the calculations shown in chapter 6.) Stock discussed in this illustration (Table 3.8) could be the same as the overall inventory shown in Fig. 3.4, where the total inventory includes the safety stock in the calculation.

Practical methods of reducing stockholding

The approach

Going back to the basic principle of

decrease in stock = output − input

shows that the way to reduce stock is to decrease input and increase output. A normal range of stock in stores comprises fast moving parts, a wealth of obsolescent and special items and fluctuating quantities of other parts, some in short supply, some overprovisioned. In this situation there is potential for decreasing stock by a significant amount, probably 30% over 18 months.

For many items there is little manoeuvrability for changing the unit cost significantly. However this should be considered, since a 10% reduction in price will cause a corresponding drop in inventory value in the long term. The average value of a stock line is given by multiplying

unit cost × average stock

The average stock quantity is typically half way between maximum and minimum[1] stock. The minimum should be the safety stock and the maximum occurs immediately after a delivery. If the stock falls to safety stock level before delivery then:

$$
\begin{aligned}
\text{average stock} &= \frac{\left(\text{max} + \text{min}\right)}{2} \\
&= \frac{\left(\text{safety stock} + \text{order quantity}\right) + \text{safety stock}}{2} \\
&= \text{safety stock} + \frac{\left(\text{order quantity}\right)}{2}.
\end{aligned}
$$

In order to reduce the average stock, therefore, there needs to be a reduction in either order quantity or safety stock.

Batch quantity

The lowest inventory will result from holding the same order size as the issues to customers, and of course this will be at a minimum if the order

---

[1] The 'minimum' on some systems is set to the review level for triggering resupply even though the stock continues to fall below this level until the delivery arrives. Here, the 'minimum' is the lowest average stock level immediately before delivery.

size is small. For those items which are individual, then the minimum order size is one. The parameter for order quantity on this basis is therefore the simple message 'buy one'. This is also the obvious answer for the majority of slow moving stock items. It is not so obvious when stated in the form 'hold the least stock cover for the most important stock items', a consequence of good stock control and also of just in time supply. Minimising the delivery quantities of the A class items (or those contributing to the highest stock value) is normally the most fruitful way of reducing stock value.

Safety stock

Minimising the safety stock could result in lower customer service. Because the balance of stock is never perfect and the levels constantly need to be changed, availability provided by the stock management needs to be monitored continuously. Therefore the first step in controlling the high value stocks is to ensure that the safety stock is sufficient to provide the service required and not more. Achieving a higher availability level than necessary is very expensive. The availability for each of the high value items has to be matched to the requirements of the market. Balancing these stocks is most important. As the safety stock depends on the availability required, the lead time and the variability of demand, the next topic to consider is lead time.

Lead times

By negotiating shorter lead times, the safety stock can be reduced. Modern purchasing practice should eliminate random demand 'over the counter' but it does require collaboration, forecasting and supply chain management so that the supplier can arrange to make items available by the demand date with a much shorter lead time. Lead time can be reduced by negotiation, and in the context of reducing stock investment, it is the few major stock items which will offer the most opportunity to negotiate.

The third factor is demand pattern. If the demand were 10 every week consistently, then there would be no need for safety stock. (There are people from some areas of business who would question this simple statement, since the holding of extra stock would enable extra orders to be serviced should they arise. This extra stock is not safety stock but an investment in opportunity, or folly, and should be considered separately.) The question to be answered is 'what opportunity is there to smooth out the demand pattern?'

Table 3.9 Stock records for aluminium suppliers

Week	1	2	3	4	5	6	7	8	9	10	11	12
Demand	225	2625	75	100	200	50	125	100	0	2050	300	125

An example

The stock records for a range of materials sold by a major supplier of aluminium were typically as in Table 3.9.

The items were aluminium circles (flat round sheets) in various sizes used for making pans. The conventional stock level calculations suggested very high stock levels in order to cope with the demand peaks in weeks 2 and 10. This introduced a high level of stock for these items and inflated the inventory value.

Investigation of this order pattern showed that across the range of products, these peaks were caused by one major customer who had an order cycle of 2 or 3 months which was causing the large demands. There are several solutions to this problem:

- *Understand demand* – Greater understanding of the demand pattern improves the forecast. Since safety stock is required to compensate for forecast inaccuracy, the improvement in forecasting reduces safety stock. There is therefore the opportunity to try to smooth customer demand by getting the customer to buy little and often.
- *Flatten demand* – Remove sporadic demand from the stock by buying to order and giving the customer a longer lead time.
- *Agree scheduled supply* – It would be better to organise a continuous supply by scheduling delivery.

The reduction project

Order quantities and safety stocks should be examined to ensure that they are synchronised, starting with the item with the most stock value and continuing down the stock value Pareto curve. A target should be set so that the stock reduction for each item can be judged against it. For example an overall stock value reduction of 20% can be achieved through a 25% reduction in stock of the top 20% of lines. Starting from the highest value stock line, a percentage reduction should be attempted for all lines. This is unlikely to be achieved for some items, but it is essential to concentrate on achieving the target for the highest value lines or the overall target will not be met. The value underachievement on some items has to be compensated by extra value savings on other lines.

Practical methods of reducing stock

Stock reduction is usually undertaken as a project and concentration should be given to achieving the objective in a short time. By examining the first few lines in the Pareto curve, major savings can be achieved on a short timescale. The next step within each area of activity is to find the major cost items. By Pareto analysis it can be assumed that 80% of the value of stock exists in 20% of the item types. Items with the most stock value are likely to be a mixture of fast movers with reasonably high unit value and high value items for which there are relatively few demands per year. Pareto analysis shows that the way to reduce the stock value is to concentrate on the high stock value items, whether they are slow moving or fast moving. The value of inventory is the same for each and the financial investment is the same. In fact it is easier to vary the inventory levels for the faster moving items. As far as basic Pareto classification is concerned, there is no difference between the stock for fast moving and slow moving items with the same stock value. The differences will be managed through the individual stock calculations for the items.

The stock of a fast moving item is likely to be consumed quickly and consistently though servicing a wide variety of demands from a large customer base. The inventory profile for a fast mover is therefore large numbers of demands taking a small proportion of the inventory. Demand is relatively stable. For high value items, there are relatively few issues to a smaller number of customers. This means that the demand is more unstable for two reasons:

- As the stock lasts longer, the risk of demand changing within the stock cover period is greater and therefore there is an enhanced risk of the stock remaining as obsolete or excess.
- The demand comes from only a few customers, if any of them reduces their offtake, this will have a much more significant effect than if there were more customers.

It is therefore especially important to manage risks on the slow moving, A class items by forecasting, ordering little and often, and collaborating with customers and suppliers. The 'A' items are selected for special control, applying a scheduling approach where possible and negotiating closely with the supplier. The supplier may be keeping a stock of these items specially or can be persuaded to do so if desirable. The delivery quantity for A class must be cut to a minimum and reviewed each time an order is placed.

Specific stock reduction can be achieved through a dedicated project. Continuous stock reduction is a background job for all inventory managers, but since insufficient focus is given to it, arranging a project tends to be more effective. The following steps may be taken:

1 • Identify the potential for reducing stock.
   • Find the total stock value.
   • Find the annual material usage.
   • Calculate the total stock cover for the various stores or categories of stock.
   • The results can then be tabulated to identify which areas of stock have the best potential for significant value reduction, under the following categories: stock type, stock value, calculated stock turn and target stock turn.
   • The analysis of stock type can be generalised for distributors. Typical classifications are: market sector, product type, supplier, or other obvious classes. For manufacturers classifications can be: raw material, work in progress, bought out parts, finished goods, consumables, etc.
   • All that is required is to reduce average stock value of significant stock lines. With this analysis it is easier to see which areas to concentrate on in a stock balancing programme. By assessing targets based on experience an initial plan can be launched, to be followed by a detailed analysis of appropriate targets at a later time.
2 The next step is to classify stock items into ABC.
3 For the few A class items, it is important that the minimum stock is held for the availability required. This is where the major effort in stock reduction should be focused. A items have to be reviewed individually and the supply arrangements reconsidered. Significant reductions are to be sought, through supply practice and better demand forecasting.
4 The reorder level and safety stock depend largely on the lead time. It is, therefore, important that lead times reflect current trading conditions and are kept low through contact with suppliers (especially for A class items).
5 For the medium price items, B class items, computer monitoring of safety stocks and order sizes will enable smaller batches to be ordered and safety stock to be reduced.
6 Avoid ordering more than (say) 3 months' supply of anything. Setting a limit ensures that the amount of slow moving stock is reduced. Very long lead time items will have several purchase orders outstanding simultaneously.
7 Low value items purchased in category C are often standard products which can be obtained off the shelf. By proper planning, these need

not be kept in stock. They can be purchased to meet demand or provided by a supplier on consignment. However, an efficient reorder procedure is necessary or else the works van will be touring the neighbourhood continuously picking up odds and ends.

8   From the stock turnover figures it becomes obvious that some items have too much stock cover and are simply contributing to stock value. They are either obsolete or have very low sales. From our introduction to this section the formula shows that this stock has to be disposed of in order to have a low inventory presence.

The first step is to decide which stock is obsolete and to remove any chance of it being reordered by marking the records accordingly. The most profitable way to dispose of stock is through the servicing department. They can often sell obsolete lines to customers with old machines at full sales value or at an offer price near to it. Stores are also likely to hold proprietary products such as components, fittings, motors, consumables, bearings and packaging which can often be sold back to the supplier at a low price. With high inflation over the past few years this return value can equal the original purchase price and so there may be no loss of assets involved.

By concentrating on major cost items the value of obsolete stock can be reduced. It generally takes time to negotiate the sale of items and therefore can be viewed as an ongoing project. Where these approaches still leave a significant number of useless items in stock, then disposal for scrap is the simplest course and there should be an approved budget for this. The loss of assets through scrapping stock reflects directly on profit margins and overzealous disposal projects can mean a period of poor financial results.

Armed with a budget for scrapping, items can be thrown out of stores and written off until the budget is used up. As stores are often short of space and control of slow movers is tedious, it is convenient to scrap a large number of the lowest value items in stock, thus reducing the number of stock lines and consequently freeing most space.

9   Design and planning departments can help to reduce stock requirements by large amounts. All they need to do is to use the same items widely rather than marginally different ones for each application.

A stock rationalisation exercise is normally fruitful for standard items such as electronic components, motors, fasteners, raw materials, packaging items, gear wheels and tools. This can be carried out initially by inventory control and then passed on to technical people to continue.

10   The order is the cause of increased stock. Therefore it is necessary
     when undertaking a stock reduction exercise to look at every order
     placed to ensure that:
     • The item is required at the time purchased
     • A minimum is ordered.
     • There is no stock existing which can be used instead.

11   To maintain overall control of the stock reduction project reliable man-
     agement information must be available on a monthly or weekly basis
     for such things as the value of items on order, stock cover and avail-
     ability. By plotting them on a graph the effect of these actions can be
     measured continuously.

4

Just in time management

The zero inventory philosophy

Conventional and just in time approaches

Stock exists because items have been bought before they are required. It is normally uncertainty or overcaution that causes stock. The principle of just in time (JIT) is simply that we have items when they are needed and none when they are not needed.

The idea may be simple but the application of JIT has given the opportunity to decimate stockholding without affecting customers. Instead of trading availability and stockholding as discussed in chapter 2, the trade-off is between organisation and stockholding. The better organised and controlled the supply chain, the less inventory is required.

Companies which are considering how JIT can work in their business, or avoiding it in a Canute-like manner, should realise that JIT is an outcome of other techniques, not a technique of its own. It is the logical aim of tight inventory control, effective process planning and plant design, workforce motivation, cost reduction, logistics and even material requirements planning (MRP). The optimisation of these together inevitably leads to the JIT approach. The elements of JIT are the techniques to be developed, for example:

- supply what is required
- supply the quality required
- reduce lead times
- organise effectiveness
- use all the expertise available (i.e. people who do the jobs plus technical backup)

These are the fundamental changes which lead to JIT – all good inventory management techniques. The most important of these improvements has

Table 4.1 Contrasts between conventional and JIT stock control

Conventional	Just in time
Satisfied with the status quo	Continuous improvement
Lead time is fixed	Reducing lead time is a continuing challenge
Product range is a sales issue	Product range reduction is an inventory issue
Management provide methods	Operators are responsible for practices
Stock in case of customer demand	Purchased to meet demand rate
Convenient purchase batch size	Buy singly or small quantities

been in quality, particularly in management processes – quality of records, supply, delivery, forecasting and target setting.

Stock controllers have been sceptical about the efficiency of frequent deliveries of smaller batches without investigating all the options and the potential. There are many instances where the accepted delivery quantity is now much smaller than it was a few years ago and is destined to continue to be reduced. Some of the contrasting features of managing conventional stock control and JIT control are shown in Table 4.1.

JIT works as a pull system whereas conventional stock control and material requirements planning are essentially push systems. In the push system the stock is provided for the next stage of supply, for example buying items for sales to sell or starting manufacture without having the total production path clear. The philosophy with push systems could often be thought of as packing so much into the warehouse or production plant that they will have to send some items out because there is only limited space!

With a pull system the first action in the chain is that the item is demanded. To satisfy this demand there is an item in stock. As soon as this stock is used up, then another item is supplied, either from outside or from a production process. In production the demand filters backwards through the processes creating a demand down the bill of materials and ending with issue of one product's worth of raw material.

There are several continuous improvement processes which are associated with JIT including product quality, process efficiency, information systems and operating value-added activities more effectively whilst eliminating non-value-added activities.

Who can use JIT?

Quality process improvement is usually thought of as a continuous journey of improvement, with no definite ending, since there is always more potential. From the point of view of material flow the principle of JIT is obviously ideal but it is often difficult to implement in practical situations unless the conditions are right. Of course the right conditions do not happen by accident and anyone looking for the benefits has to work hard to create the appropriate situation. JIT supply should be considered as a quality process, although most of the objections to JIT are based on lack of quality in some aspect of supply or demand. If stock levels are incorrect, this is often the result of complacency or lack of understanding (usually both!). There is no perfect solution to stockholding, but like any other quality improvement process, JIT operations gradually develop an existing unsatisfactory situation into an improved one. A decrease in lead times and simplification of processes should be the aim of all inventory managers. From the viewpoint of JIT, time is a value-added commodity and wasting it is unprofitable. The more time saved the better, and continuous improvement means reducing the timescales.

The definition of JIT presented so far can apply to any material management process which actively minimises timescales. The purist would think that there is more to JIT than this simple concept and there are some specific concepts for achieving this reduction in timescales, particularly:

- Desire to improve
- Simplification
- Demand led supply (pull)
- Quality conformance
- Devolution of responsibility

If these concepts are put into practice, then an operation has a JIT philosophy (most of these concepts are fundamental to good inventory management anyway). Supported by improvements in communications and driven by the need for better service and lower costs, the influence of JIT has been felt in all types of business and has fuelled change.

JIT in manufacture and warehousing

JIT was taken up by some sectors of manufacturing originally as a space efficient method of production. For some supply chains, the application of JIT is natural because of the nature of the products or the processes, for example in fast cycle manufacture, where a stage in the production process takes very little time. In a high volume manufacture, delays between processes would cause large stockholding (or more specifically work in progress). In other businesses JIT techniques are less easy to apply. Retail shops find it important to hold stock so that the customer can have a choice

Table 4.2 Warehouse stockholding

Stock cover	Inventory investment ($)
52 weeks	2500
10.4 weeks	500
6 weeks	288
4.2 weeks	202
1 week	48
1 day	10
1 hour	1.3

of several options, so JIT supply would not be appropriate for this stage of supply without changing the nature of the relationship. If the customer were buying the same item by mail order, then there is a good opportunity to use JIT techniques.

Although the original impetus for JIT was for space saving, the greatest benefits are financial. In chapter 1 (section on *Objectives for inventory control*) the benefits of lower stock on profits were illustrated. This magnitude of improvement is likely to occur for a second time as a result of applying JIT.

Consider the example used in chapter 1 with each of the companies having a $5 million turnover. Often companies find that half their costs are in purchased materials, let us assume that the same is true for these companies.

If these companies are distributors, then the stock cover for Slack & Co is 52 weeks and for Tight Ltd. is 10.4 weeks. If the application of JIT on selected lines (product types) could reduce the overall stock cover to 1 week, then this would result in an inventory holding of only $48 000, a saving of $202 000 even for Tight Ltd. The stock cover and relative inventory investment are shown in Table 4.2 for a turnover of $5 million. The decrease in inventory values are large when the initial stock cover is high. However as stock cover becomes low, the potential savings are lower and the risks greater. It is then not the value of inventory that is the driving factor toward zero inventory, but the operational cost savings.

Applying the same logic to a manufacturing situation shows additional benefits through the savings in planning resource. The operational situation is more complex in manufacturing. The material cost starts out as say 50% of turnover but then collects added value until it is worth typically 70% of sales value. For simplicity let us assume that there is an equal balance of inventory throughout manufacture, so that the average inventory value is 60% of sales value.

Table 4.3 Manufacturing inventory

Throughput time (weeks)	Inventory investment ($)	Number of jobs in progress
43.3 weeks	2500	2167
8.7 weeks	500	433
6 weeks	346	300
4.2 weeks	242	210
1 week	58	50
1 day	12	10
1 hour	2	1

A further aspect of manufacturing is the number of jobs in progress. It is typical for a company with a turnover of $5 million to issue about 50 new jobs to production each week. (The logic of the following discussion works just as well for companies with different numbers of jobs and turnover values.) Using this figure a manufacturer's performance table can be created as shown in Table 4.3. Stock cover in manufacturing is required for the duration of average throughput time. For Slack & Co., the $2.5m inventory equates to 43.3 weeks' cover and a total of 2167 jobs in work (assuming that all the inventory value is in work in progress, WIP). For Tight Ltd. the WIP equates to 8.7 weeks' cover and 433 jobs.

The advantage of reduced lead times in manufacturing is twofold, as can be seen from Table 4.3. Inventory investment is reduced as well as the number of jobs and the complexity of the materials management. It takes a great deal of effort to control 433 jobs, but much less to track 50.

The illustration shows that major cost savings go hand in hand with reductions in manufacturing complexity. Inventory cost savings in particular can be achieved by cutting the lead time down to a few weeks, days or hours, and then benefits flow from other operating costs, such as less shop floor space required, improved flexibility, less expediting, more control and less material handling. There is a similar simplification in warehousing where lower stocks means less to count and less to control.

JIT environment

The ideal situation

Over the years the perception of lead time has changed in most businesses. Lead times have shortened due to the combined influence of customers and accountants, the former wanting faster service and the latter lower inventory. This has been a gradual culture change, but the move to true JIT requires a further leap in thinking. An attitude of 'we're doing alright, don't risk it!', has to be replaced by 'this is what we need to do, now how do we make it happen', that is, continuous development. For JIT individuals have to create an environment where things do not go wrong and their positive attitude is the basis of JIT philosophy, whose essence is reliability and consistency. Ideally this reliability must pervade the whole of the material control operation, not just one stage. Performance is dependent upon a number of factors, for example suppliers, the market, the company itself and the individuals within it, and these factors will always throw up a problem of some sort. (These should be considered as the challenges of every day stock control.)

In high volume manufacture it is easy to see how items can be transferred from operation to operation, and lead times and cycle times can be kept very low. In distribution operations where the demand is variable and uncertain, the application of JIT is rather more difficult, but not impossible.

Typical features of the ideal company where JIT concepts could be applied would be:

- Narrow product range
- Manufacturer
- High volume
- Stable market
- Influential
- Good quality management
- Local, reliable suppliers
- Dependent suppliers
- Fast cycle processes
- Affluent.

The one major element which was left out of this ideal list is personal commitment, and although comparison with one's own current business profile might prompt many to discard JIT as inappropriate, a positive attitude will prompt others to consider how close their business is to these ideals and how JIT can work in reality. This is a major factor because successful JIT depends on commitment by a core of people at operational level and equal dedication of managers and directors. As in all good management, it is useful to have monitors for success. These are local targets set up by the local team to set and manage their own performance and criteria for success. Quality and throughput will be part of these measures.

Repetitive processes

JIT can be seen to work best in repetitive processes, whether in warehousing or in manufacturing. Of course, almost all of business is based on continuity of supply, so this is not really a major constraint. For a usage of 20 per month, a traditional approach could be to buy 20 at a time, whereas JIT would require one per day. Such a small order rate may not appear economic for one item, but it would be when dealing with a large number of items. Consider a packing or manufacturing line that each week completes:

• 2000 of item A
• 3000 of item B
• 4000 of item C
• 2500 of item D
• 1000 of item E

Instead of working on each type of item in turn, it would be better to provide each item at the rate at which they are required. This can be achieved in two ways, (i) first to produce at a balanced hourly rate across the whole range and (ii) second to produce to order and adjust the rate to suit demand. For the first option the rate of production is consistent. The supply rate is shown in Table 4.4. For this option to work, there has to be a consistent demand or a buffer stock somewhere in the system. This is generally the way that JIT operates in warehousing and manufacture where internal planning is on a fast cycle basis, but customer demand is on a slower cycle basis so that demand fluctuations can be smoothed out over the customer supply cycle.

Examples of this situation are the automotive industry and high technology equipment manufacture. In these cases the supply can be geared to a constant rate of output and the resources organised to efficient and cost effective supply (and production where appropriate). In Table 4.4 the options for the stores or production unit are to have a single throughput channel which completes an item every 11 s, or to have a multiple of channels which deal with a single or a limited range of items. In the automotive industry one production line generally implements a variety of processes at the same time but on one car at a time (with a parallel line for flexibility and to offset the effects of changeover or plant failure). In a mail order warehouse the capital investment per line is not so great and stock items are more diverse, so that generally there is a number of specialised packing lines.

In warehousing it is important to apply JIT where it is easiest to use, and

Table 4.4 Continuous supply cycle times

	Throughput per week	Throughput per day	Cycle Time (s)
Item A	2 000	400	68
Item B	3 000	600	45
Item C	4 000	800	34
Item D	2 500	500	54
Item E	1 000	200	135
Total	12 500	2500	11

the conditions discussed earlier in this chapter point toward the A class items with high usage, more consistent demand patterns and significant benefits from inventory reduction. The items sold can be split into two types, those controlled conventionally and those using JIT. The major items will be JIT controlled and the wide variety of low turnover value items controlled conventionally.

Having carried out a full discussion of the first option (i), the control for the second option (ii) becomes an extension of this. Here the customer requires items on a fast cycle basis. This means that fluctuations in demand have to be taken up by varying the capacity available. This is less efficient in the use of resources but may avoid inventory buildup elsewhere in the supply chain. This option includes the full 'pull' concept of MRP as discussed later in the section on *Pull systems*.

JIT supply

Processes can be considered to be suitable for JIT if the timescale is in hours rather than days. For most suppliers this requires special arrangements for feedback of demand and issue of items within a short timescale. Effective JIT supply requires the same approach as that used for single source suppliers (see chapter 8): there has to be a longer term agreement on anticipated throughput levels, quality management and commitment by the supplier to provide an infallible supply of specified items. The supplier should be treated like an intrinsic part of the organisation.

One of the changes in attitude that this brings is the approach to transport. Often the suppliers organise delivery so that they incur the cost. With just in time supply there is an advantage if the customer collects the items at a time convenient to them (the pull system being extended in this way through the transport system), since the transport cost would be incurred in the final cost whichever company pays for it.

For a fast pull-type relationship to work, the supplier has to be local to the customer, otherwise the reactivity is not good enough. Location is of significant importance and some suppliers have moved to the vicinity of a major customer in order to ensure close cooperation.

JIT requires more frequent deliveries from suppliers but this does not automatically lead to higher transport costs, although it could do so if no changes are made to mode and arrangement of supply. Simple illustrations of where there is definitely no extra transport cost are:

- When the freight is charged simply by weight.
- When a wider variety of items is put into each shipment and these compensate for the reduced amount of bulk orders.
- Where freight is charged at a flat contract rate.

Costs might increase:

- Where the same delivery method is used more frequently.
- Where batches of items are bought separately from different suppliers.
- Where a flat rate per consignment is charged.
- Where transport is arranged one journey at a time.

JIT philosophy requires close cooperation with suppliers of distribution services as well as of goods. By moving toward longer term, higher volume commitments with suppliers', prices can be reduced based on the total traffic volume over an extended period. The distributor requires a continuity of demand and the opportunity to work out routes, loads and schedules well ahead and this will reduce the operating costs as long as the business is significant for the distribution partner.

Many people are sceptical about increasing the number of deliveries without increasing the overall costs. Others who have actually tried increasing the delivery frequency have found distribution partners who can provide the service without increasing the overall cost and there are an increasing number of companies working in this way.

The general cost savings produced by JIT are discussed below. The economics of frequent deliveries can be added to these savings for delivery processes. Some of the typical benefits are:

- Smaller delivery quantity means smaller loads, which enable different transport methods to be used (parcel post instead of lorry load). Small bulk transport often costs considerably less, since the operating costs are reduced.

- Frequent fixed route deliveries enable a carrier to have a base load upon which to build other business.
- Frequent deliveries do not require planning and management effort once set up.
- Discounts on contracts can be significant.
- Standard pack quantities and containers are smaller, less expensive and recycle is faster.

The secret of success in saving costs on frequent JIT deliveries is to:

- Plan carefully so that delivery schedules are not altered at the last minute (this is easier with JIT since the planning horizon is relatively short).
- Be bold – radical changes in delivery methods are required if costs are to be minimised (this often means a complete change in transport method).
- Negotiate carefully – transport costs are adjustable because they contain a large fixed element of cost. The apportionment of this cost (e.g. running a lorry from A to B) can be carried out in any way. Transport suppliers may be persuaded to apportion the costs away from routine major deliveries.
- Be open – the size frequency and costs of delivery required should be shared with the transporter. Their relevant costs should also be known in order to ensure that their service is viable.

Quality management

Concurrent with the development of JIT there has been a great move to improve quality. This has taken two forms.

First there was the move toward consistency through the development of quality standards (including ISO 9000) which were instrumental in providing reliable performance, not only for the products provided, but also for procedures and paperwork. These standards have changed attitudes within industry and there is now an expectation that items provided shall be correct to specification.

The second form of quality improvement was the development of the total quality management (TQM) philosophy, which does not accept current quality and strives always to provide gradually improving quality. When this philosophy is applied to the operating processes of the organisation, the key concerns are the elimination of waste and operating effectiveness. TQM then becomes a practical approach to productivity improvement, standardisation of packaging, improvement of the quality of delivery on time, record accuracy information and operating effectiveness and the complete operating processes of the company.

Quality management principles integrate well into stock control methods and operations because it is a basic assumption of stock control systems that the items supplied are fit for their purpose. If a delivery batch is not acceptable, it has to be replaced immediately for the stock control calculations to be correct. Delivery in smaller batch quantities often reduces the overall effect of rejected supply since it impacts on fewer customers' orders and leaves less time before the next delivery. Non-conformances will cause extra work, especially:

- Creation and holding of extra records as a cross check
- Quality inspection on goods inward (also causing booking delays)
- Sales requiring physical stock checks rather than relying on the recorded quantities
- Time spent investigating and checking
- Arranging replacements and priority deliveries.

Formal specification of inventory management procedures enables them to be reviewed and identified for quality management processes. As time is costly, the allocation of time to administration has to be done sparingly. The cost and benefit accrued by maintaining full records and checking procedures has to be balanced against their added value benefits. Non-essential activities have to be avoided (such as paperwork and manual systems, where there are computer records). Procedures have to be improved and rationalised continuously, to maintain the impetus of the quality improvement.

Customer support

Supplier quality is not only measured by the acceptability of the product, but also by the customer support provided. Rating of suppliers should include:

- Delivery performance
- Packaging and labelling
- Information systems and communication
- Support
- Style
- Flexibility.

Supplier quality issues are discussed fully in chapter 8.

Advantages of JIT

Operational benefits

How can it be more efficient to deliver in small quantities, manufacture in small batches and increase the amount of administration? These are often the queries from those brought up in the traditional school of thinking. Equally someone who only knows JIT would ask:

- Why do you buy things when you don't need them?
- How do you know what the demand is so far ahead?
- What is a warehouse?

The accounts of a conventional company show a large investment in inventory. In the case of the two companies discussed in chapter 1, the profits were increased by a low stock philosophy. If M Tight Ltd. were to adopt a JIT philosophy this could bring the stockholding down even further to one or two days' worth, say $20 000 total stock value. This would release a further $480 000 cash for the business and increase the return on assets from 16.7% to 20% as a minimum. (As a move to JIT will eliminate the stores and some overheads, the real improvement will be greater than calculated here.) In this case each company had a large investment in fixed assets. For a distributor with rented warehousing the increases in profitability are very large, even if the average stock value turns out to be weeks' worth rather than days' worth.

The operational benefits arising from JIT are:

- Inventory investment
- 'Supply to order' instead of 'provision for stock' (see chapter 7, section on *Managing lead times*)
- Easier forecasting giving less slow moving stock
- Better flexibility
- Simplified administration
- Waste elimination
- Less waste should there be a problem

For each of these operational benefits there is a corresponding cost benefit which can be offset against any additional costs which arise. These additional costs usually occur because methods have not been changed to suit JIT. For instance if an item is delivered in a batch once per month, it can be invoiced, delivery documentation produced and payment made. If through a change to JIT the item is delivered every day, then it would not be sensible to place a purchase order for each load, to raise delivery paperwork and to arrange separate payments for each load. Information is still required for control, but the information system has to be recreated to meet the new conditions. In the short term this may not be possible and so extra costs can arise. As JIT embodies the process of continuous improvement, the inefficiency will eventually be eliminated.

Efficiency benefits

JIT supports the continuous output of a variety of different items, which in turn creates a variation of items going through the processes (picking, manufacture, packing, etc.). Flexibility is essential, but conventional thinking is that changeovers are to be avoided since they take up time and use resources. The JIT approach is to find the output rates as illustrated in Table 4.4 and then determine how to achieve the throughput and mix. Changeover time is therefore considered as a variable and this is a basic difference in the JIT approach. Normally the changeover time is associated with a particular product. Conventional logic is that if it takes 10 min to changeover for a particular item type, there is no point in making just one (which takes 15 sec), better to make at least 80 to be efficient. Accountants usually back up this type of simplistic logic with financial break-even analysis (see for example economic order quantity (EOQ) in chapter 8, section on *The EOQ approximation*). This results in large batches being produced and high stocks being held. The real situation is that changeover time is a variable depending on what the change is from and to. Fast changeovers give the opportunity for lower stockholding. On a packing line, the change from one type of item can be fast if they are packed in the same type of box, but slow if the box size has to be changed. The secret of success is therefore always to run the right job sequence so that the number of short changeovers will be large and long changeovers small. This requires planning and load balancing which is difficult when forecasting for weeks ahead, but easier when looking ahead a few hours, as required by JIT.

Assuming that the items in Table 4.4 are all processed on the same process line, then one item needs to be completed every 11 sec (5.6 per min). If the process is carried out on a cyclic basis as shown in Table 4.5, then the cycle is repeated every 4.5 min. The number of each items in the cycle is shown, together with the output rate required to cope with the average demand. If the 4.5 min cycle is to be carried out most efficiently, the changeover times between the items need to be reduced. There are two ways in which the changeover time can be reduced, namely by improving the planning of production sequences and by reducing the time per changeover.

Table 4.5 Cyclic production

Line	Throughput per min	Number per cycle
Item A	0.9	4
Item B	1.3	6
Item C	1.8	8
Item D	1.1	5
Item E	0.4	2
Total	5.6	25

Time per complete cycle = 4.5 min

Considering these two options, improvement in product planning of sequences can make most changeovers minor. Within a short repetitive cycle it is much easier to sequence the production to optimise the setup time than it would be with conventional priority planning. The quantity of each item can be varied to suit the demand required if this varies, and items with the same setup can be substituted should this be required by the programme.

Changing the production philosophy from weekly batch quantities (see Table 4.4) to cyclic JIT production (Table 4.5) will reduce stock levels by a factor of 20, and also save overall cost. JIT organisations often have small management teams with few management levels. This is a result of the JIT process itself. The planning operation is simple and the management is largely a repetitive activity which does not require a great deal of managing, especially as the direct process control is carried out by the direct operators.

From an accountancy viewpoint, if the cost of supply is split under the headings of materials, labour and overhead, the major impact can be in the lower overheads resulting from the simpler operations. The proportion of overheads applied to the direct costs is much lower in JIT operations.

Stock control using JIT

Lead time reduction

Inventory control has to work under the conditions imposed by the market. To get the stock levels right calculations are based on the real information which is available and not on idealised values. There is no reason why the situation should not be improved because JIT gives potential for improvement, particularly the potential to shorten supply lead times. In working with a dedicated supplier, effective lead times can be changed. The supplier and customer can work together and share the risk.

Lead times are a result of queuing, distribution and sometimes process time (see chapter 7). For JIT suppliers, the demand is assured by the agreement with the customers, but the demand timing and quantity varies. As the JIT demand is frequent, the effects of these variations cause only minor stockholding and delays – days' worth not weeks' worth. Small stocks enable the supplier to meet the small daily fluctuations created by the customer very rapidly. Lead times can therefore be reduced.

To set up an effective JIT supply

- The customer must *smooth the forecast.*
- The supplier needs a regular demand for the item in order to *estimate the average demand rate.*
- The supplier also needs a good estimate of the long term demand for their planning in order to *understand the supplier's constraints.*

Suppliers can extend some aspects of their service and struggle to meet other aspects.

- Greater understanding improves the match between the needs of the two companies and *builds confidence.*
- Trust is the major ingredient for effective collaboration in order to *maximise supplier resources.*
- If suppliers know the volume of demand they can react much more quickly to individual requirements. For instance, a supplier of painted items could often paint and deliver within a few hours, as long as the demand for the item is expected and the choice of paint colours has been agreed previously. This enables them to offer *flexibility in demand changes.*
- As short term demand is always variable, an agreement is needed to ensure the extent to which demand can be raised or reduced. This avoids failures in the JIT supply. It is therefore necessary to *communicate and monitor.*

Close liaison between the two parties involved is essential. It involves regular meetings for assessment and review between people at operational level.

Reducing the lead time is a continuous process starting with the A class items. These have the greatest cost benefit and are also amongst the simplest to change. They have most importance to the businesses and are amongst the easiest to forecast.

Pull systems

The essential difference between the JIT approach to material control and the stock planning approach is in the mechanism for triggering replenishment. Conventional inventory control predicts demand, works out how much stock is required to supply it and then procures that quantity. JIT operations only acquire stock as a result of demand. This can be thought of as a 'pull' process, with the issue of an item triggering the demand for another – the number of triggers depends upon the rate at which demands are received. Stock level systems purchase items to cover forecast demand (in larger quantities) and then try to supply the items to the customers (a 'push' system). The advantage of the pull system is that only sufficient stock is held to meet the immediate demand for a few hours or days. The effective supply lead time is arranged to be short (JIT) and the deliveries frequent, so therefore the stock can be very low.

In inventory management, simplicity of operation is the key to success. Kanban is an operational method, consisting simply of a ticket, which achieves a pull system from within the company or from a supplier, consisting simply of a ticket. When an item is demanded a request must be passed back to the source to provide another item. This information can be a Kanban card or a simple signalling system. The card will identify the item, the quantity required (ideally one, but possibly more) and where it is required. It will also state times so that performance can be monitored. This ticket or traveller has to be provided for the source rapidly if the supply is to be prompt. The ticket is held with the physical goods and is an easily maintained and simple method of informing the previous stage of supply. Using the principle of the ticket, there are other trigger mechanisms which can be used instead, such as coloured lights or sound. Fax or network communications can be used for external supply. The way information is transferred is immaterial as long as the process is instantaneous, completely reliable and accurate.

An obvious application of JIT is for manufacturers producing large quantities of similar products. They can set up a flow line and have in-line stocks of a single item or a small batch between each process stage. The trigger for manufacture is simply 'use one so make one' starting from the last process and feeding back gradually to the first. The restructuring of production into a flowline is a most effective way to manufacture. Where the product range is wider and the process routes differ, cellular production is more applicable, and the pull system is used between the cells and by the cell operators for processes within the cells.

For purchased items the supply is triggered in the same way but with larger batches. The delivery batch would be expected to meet the demands for a few hours or a day. As the lead time is normally over an hour, a supplier kanban does not run out before signalling. It has a residual stock to cover the demand during the supply lead time.

JIT has made a major impact on manufacturing and the reduction of lead times is changing the whole of inventory control. The changes will provide competitive advantage for those companies with the best ongoing developments of JIT.

5

Organisation and management

Where stock control fits into the organisation

Nowadays the inventory controller has extra responsibilities. The job has grown from store keeping to become a management discipline of maintaining the inventory to meet company policy. Gradually this responsibility has broadened to include all stock whether inside or outside the stores. The place of inventory management in the organisation has altered as responsibilities have changed. This is illustrated in Fig. 5.1. The original role of stock control was part of the physical stores control shown in structure A and its position in an organisation was lowly, working for either purchasing, finance, or in manufacturing companies, for production. In this structure, stock acquisition was initiated by buying and sales pressure and there was no clear responsibility for, or focus on, inventory control. The tightening of company objectives changed the approach and separated stores and stock control activities. Inventory management is now accepted as an operational activity and works alongside purchasing, as illustrated in Fig. 5.1B.

This arrangement has gradually evolved into structure C as pressure increases for better performance. The central role of inventory management has been recognised in many businesses, both in distribution and manufacturing, and material management or logistics directors have been appointed to coordinate various aspects of inventory management into one supply activity. Stock control encompasses:

- Purchasing, especially reordering (but see chapter 8)
- Forecasting demand
- Planning inventory against company targets
- Stock allocation and delivery promising
- Monitoring and controlling service and inventory and
- Production planning and control (for manufacturing).

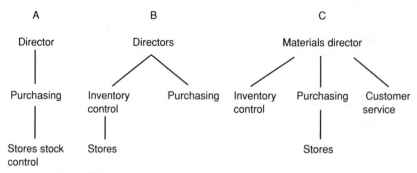

Figure 5.1 Stock control in the management structure.

These are organisational activities. The physical management of inventory, including stores, despatch and movement is then treated as a separate problem to be managed by a parallel organisation. A similar relationship can arise in manufacturing between the production control department, who carry out the planning, and production supervision itself, which seeks to work to the plans provided by the production control department.

Where there is a separation of the stock and stock control activities, the warehouse becomes responsible not only for managing the movement and holding of stock, but also for providing information for the rest of the company. It is particularly important for inventory management that the information provided is accurate. Checking stock physically is time consuming and the recording systems should be sufficiently accurate to make this unnecessary apart from stock auditing.

Stock control in a professional organisation can be carried out at a different location from the stores. Thus stock controllers can manage site stocks or multilocation storage as long as there is real time communication with the stores. If these activities are split, warehousing can be located at the most economic location for transport and customers and inventory control can be located independently at the most convenient place for management.

Inventory management is in the process of evolution. Targets are being expanded to include extra aspects of operations, cost and service and there is much potential for development (see chapter 13).

Responsibilities and targets

Rationalizing the problem

Inventory controllers have to take a simplified view of the stock items and treat the majority of them in the same way. In driving a car there is a distinction between the driver (who may not be able to repair the car) and the mechanic (who may not be able to drive). In stock control, most people are the drivers but the stock control vehicle is not well designed, mechanics are few and so the driver is often forced to make improvements. The driver who understands the mechanics can always get better performance out of a vehicle.

Every item has its own characteristics, often with a unique story – 'do you remember when the supplier's warehouse burned down?' 'The customer was frozen up for 5 weeks?' 'It was put on a boat to Antarctica by mistake?' In the past individuals managed stock through their memory and understanding but this is not enough to meet the tighter targets of modern inventory management even though such understanding is an important additional tool to use for A class items (and a waste of time for C class!).

The stock controller should largely ignore the details of the item and consider the characteristics of the demand pattern and the supply, despite the widely different types, sizes, shapes, usages, origins and characteristics of the lines, but whatever the item is, inventory control remains the same. An arguable point? What about liquids, explosives, perishable foodstuffs, pine trees? They are all different. True, but despite the differences, their supply and demand characteristics are not that different: they depend on market conditions and not on the product. Just as financial controls are applicable across different industries, so is inventory management. The characteristics of the stock determine the mix of inventory techniques to use, but the components are the same:

- Demand forecasting
- Determining safety stocks
- Negotiating supply patterns
- Avoiding slow moving stock

Inventory techniques are common for all businesses and their skilful mixing for aggregate and individual item inventory control allows many different types of stock to be managed. Simplicity is the key, but this is not the simplicity of systems which give superficial answers. It is the simplicity of operating processes resulting from rationalization of procedure, systems and approach. It arises from, say, using the same system for placing all orders and controlling all the stock; in having the same item identification throughout the business; in treating priorities through an organised system. The elements of the simple procedures have to be supported by complex tools such as software packages for forecasting, modelling, communicating, but these need not concern the daily operation as long as they work correctly. The actual logic of such tools has to be intricate to cope with the complicated nature of inventory planning.

Simplification should be considered as an ongoing challenge for development of good stock keeping. Areas which can usually be improved:

- Single stores philosophy
- Inventory system applied to everything
- Single operating procedure
- Paperless information systems
- Purchasing as part of inventory management
- Real time systems.

By rationalising these activities the professional stock controller can free time to work on the most beneficial topics and improve the level of control.

External objectives

The initial discussion earlier in this book was about setting inventory targets for availability, inventory value and operating costs. The balance between these targets is a matter of inventory policy. The management of inventory relies upon routine and regular monitoring of these values. The objectives set for inventory management should be chosen to meet the immediate demands of the business, specifically:

- Customer service
  - measured as stock availability or delivery on time, percentage against own and customers' targets
  - Analysis of age distribution of lateness
  - Assessment of customer satisfaction – survey results
- Inventory investment
  - Value
  - Stock cover ratio
- Operating costs
  - Warehousing and inventory operations costs
  - Cost per transaction, movement and purchase.

The process of setting targets usually begins by measuring current success. Current levels of achievement are assessed to see whether they are good or bad. (If inventory control staff consider that they have the correct stock level, this is often because of their complacency and not because of adequate achievement assessment.) Targets can then be set through negotiation with those who are affected by them and with those who have to achieve them. The three target dimensions are:

- Quantity – how much the monitored values will change.
- Date – the stages by which the quantities will be changed.
- Responsibility – who will be accountable for achieving the results?

The prime aim is to meet all the inventory targets simultaneously and not to sacrifice one to meet another. Measurement methods should meet the prime business needs (in chapter 2 alternative ways of measuring customer service were discussed) and should be chosen to give the most satisfactory results.

Routine monitoring of these items is required on a weekly or monthly basis. The results, both operational and financial, should be available within three days of the period end, otherwise the information cannot be used for control (although it is commonly used for apportioning blame!).

Managers should also be aware that targets are set as part of company policy and the art of correct management is to set targets which are both achievable and challenging. They should reflect an expectation of performance, not an achievement under ideal conditions. This can be difficult when trying to coordinate planning with other departments. A good method of improving the accuracy of planning is to give two targets, one optimistic and one pessimistic, and to ensure that performance lies between the two. By adjusting the two limits, a target range can be identified which is both attainable consistently and which is acceptable for planning and coordination with other activities.

Internal targets

External targets have to be supported by more detailed objectives within an inventory management department and these should include personal objectives for individuals which all contribute to departmental targets.

Customer service can be structured by setting availability targets for different types of items. Some service is likely to be ex-stock with different availability levels whilst other service will be supplied on a range of lead times. Information required to give fast direct performance measures includes:

- Number of lines out of stock
- Number of customers and orders affected by shortages or late deliveries
- Analysis of shortage according to cause
- Monitoring of customer complaints
- Details of customer care initiatives

Stock levels will differ for these product ranges and have different expectations in terms of stock turns. Items with a JIT (just in time) type of operation have very good stock turns whilst slow moving items such as service parts can inevitably achieve only relatively poor results. In addition to the inventory value, the investment can be controlled by managing purchase commitment and write off value as discussed below.

Operating costs are increasingly becoming the responsibility of individual inventory controllers. The elements of cost are controlled through cost analyses and budgets and monitoring operational activity levels often provides the basis for cost reduction. These controls include:

- Value of transactions or number of transaction per day (e.g. customer orders processed, stock reconciliations done, adjustments to stock levels, etc.)
- Processing delays or queue lengths (e.g. allocated customer orders)
- Cost per transaction by type
- Performance against budget for each activity.

Other targets are required to complete the management of inventory activities. Prime external targets do not necessarily cover all the requirements for good inventory management and there are additional aspects which are important as well. There are requirements for individuals to manage change and control projects; there are other business objectives to be met, including quality management, credibility, legal, motivational, health and safety, communication and process development. The real requirements for inventory control are numerous and, where possible, targets are needed to evaluate and monitor these objectives. Inventory management must also meet general business objectives. Targets need to be set in the areas of:

- Personnel development planning
- Improvements in systems and procedures
- Quality development for products, service and administration
- Management of change processes
- Supplier audit
- Environmental issues
- Presentation, style and image of the department

For well managed inventory control, all the requirements of the market, the operating environment and the supply chain have to be considered. Greater understanding of the real objectives will improve the effectiveness of the business as a whole and the worth of inventory management within the business.

Purchase commitment

One way of maintaining control over inventory value is to balance the values of input and output. If the value of input is kept to below the value of output, then stock reduces (as discussed in chapter 3, section on *Practical methods of reducing stock*). The value of supplier deliveries should be less than the forecast value of customer despatches in each time period. Supplier deliveries are a result of purchase order and schedule commitment, so it is the ordering process which has to be kept under budgetary control. The control process relies on collecting information on:

- The forecast value of demand for each time period (week or month).
- The amount of schedule or order commitment already made for delivery in each time period.
- The typical value of last minute (emergency) orders which are committed for each time period.

A policy on stock reduction, or increase, has also to be defined in terms of what value change in inventory level is considered practical or desirable in each time period. From this information it is easy to calculate:

1 The maximum delivery value for each time period (= forecast customer demand – value of stock reduction target per period).
2 The maximum value of order commitment for delivery per period (= maximum delivery book value – typical value of emergency orders).
3 The maximum value available for new orders placed for delivery during that period (= maximum value of order book commitment – amount of order book commitment already made).

This maximum value available for new orders is the ceiling for total order value for delivery in the time period. If this value is exceeded by placing a further supply order, then that order should be rescheduled into a later time period (or another order delayed) or else the resulting stock value will increase.

Taking a typical example, the purchase value of goods on sale each month at M Tight Ltd is $210 000 and the stock value is $500 000 (see example in chapter 1). Management want to reduce the value of stock by $120 000 over the next year ($10 000 per month). The maximum allowable delivery value per month is therefore $210 000 – $10 000 = $200 000 (step 1 above). The typical value of deliveries which result from last minute purchases is also, in this case, $10 000 per month. The amount left over for planned delivery is therefore $200 000 – $10 000 = $190 000 (step 2 above).

Table 5.1 Order book values

Period	Supplier deliveries already ordered	Value of order book available to place
Next month	185 000	5 000
1 Month ahead	130 000	60 000
2 Months ahead	60 000	130 000

Table 5.2 Inventory management targets – stock value

Location	This week/month			Last week/month			Target	
	Inventory value	Slow movers value	Weeks on hand	Inventory value	Slow movers value	Weeks on hand	Inventory value	Weeks on hand
Central								
NW store								
SW store								
NE store								
SW store								

Table 5.3 Inventory management targets – availability

Location	This week/month	Last week/month	Target	
	Availability (%)	Availability (%)	Availability (%)	Date
Central				
NW store				
SW store				
NE store				
SW store				

The current order book value with suppliers for delivery over the next 3 months is as shown in Table 5.1, and the maximum order book value for delivery in each time period is calculated in the last column. The task for the inventory controller in M Tight Ltd. is to ensure that further purchases are placed so that the planned deliveries next month are not more than $5000 and then $60 000 for one month ahead and $130 000 for the month after that.

By monitoring delivery book value in real time, an inventory manager can maintain the stock within the required budget. The control of

Table 5.4 Inventory management targets – supply

Purchasing activity	This week/month	Last week/month	Target	
	Value	Value	Value	% Success
Orders placed				
Goods received				
Late deliveries				
Emergency orders				
No of orders	Qty	Qty		

Table 5.5 Inventory management targets – outstanding demand

Location	This week/month		Last week/month		Target	
	Overdue	More than x days	Overdue	More than x days	Overdue	More than x days
Central						
NW store						
SE store						
NE store						
SW store						

stock reduction using purchase commitment is illustrated in Tables 5.2–5.5.

Inventory valuation

Unit cost of stock lines

Accounting procedures for stock are usually organised by the accounting functions of a company and are not of great relevance to the inventory controller. However, it is useful to know what these alternatives are and how they affect inventory. The options which are commonly used are explained below.

First in first out (FIFO)

Stock is valued at its purchase value. The oldest stock is assumed to be used first (as is required by good inventory practice) and the stock value is therefore the total of the most recent purchases. FIFO is best used as an accounting procedure, but not for identifying which items in stores to pick. Stores stock rotation should be arranged through the warehousing control system.

Last in first out (LIFO)

The issues are valued at the most recent purchase price, leaving the remaining inventory valued at a previous (generally lower) value. The effect is to minimise the profit on the stock being sold and to minimise the value of the remaining inventory. This method of valuation can be used to reduce the profit reported by a company and decrease the valuation of the stock. It is, however, only a valuation technique and not appropriate for organising stock movements. For example, there is a stock of 10 of an item valued at $80 each, and a new batch of five has been bought at $120. The stock value is therefore $1400. If a customer buys two, then the stock will be reduced to 13. By LIFO these two issues will be valued at $120 each, and the new stock value for the 13 items will be $1160. If the items sell at $150 each, the recorded gross profit is $60. (Using FIFO, the remaining stock of 13 would be valued at $1240 and the gross profit would be $140.)

Replacement value

Stock is valued at the current price for buying replenishments. This enables a simple rule to be applied to sales price, (such as 'sell at purchase price + 50%'). It increases the value of stockholding but requires market information to manage.

Standard cost

Fixed item value is calculated financially and held the same for a long time, usually a year, giving a good stable valuation. Differences from the standard cost are considered as good or bad variances. Standard costs are used widely in manufacturing, as they can account for typical costs and overheads.

Average value

A running average is the safest stock valuation method. The cost of new deliveries is added to the total stock value and the total value spread over the new total stock. For example, there is a stock of 20 of an item with a unit cost of $10. Ten more are purchased at a cost of $13 each. The total value is now $20 \times \$10 + 10 \times \$13 = \$330$. The stock is now 30, so the average cost is $\$330/30 = \$11$. This is the new average cost and any demand will reduce the stock value by \$11 per item.

SKU

This is a stockkeeping unit and it is especially useful in distributed warehousing. SKUs acknowledge the fact that the stock value can depend on where the item is held. If a customer requires an item which can be supplied either from the main stores or from a satellite store and the cost of distribution is \$15 from either source, where should the supply be made from? Since the stock in the satellite store has already been transported once (probably at a cost of \$15), then it would be cheaper to supply from the main store since the satellite stock has already cost more. SKUs are stock at a specified location taking into account the transport and purchase costs. It is normal to talk about a 3000 SKU warehouse, which is equivalent to saying that there are 3000 lines in that remote location. The actual stock value element of cost in an SKU could be calculated by any of the preceding methods.

These alternative valuation methods are illustrated by an example in Table 5.6. The 'cost of the next item to be issued' is the unit value of the item remaining in stock which will be issued next to fulfil customer demand. Each of the valuations gives slightly different unit costs. For LIFO and FIFO methods, allocation of the costs depends on working out from which supplier delivery the next sales item should be taken.

Aggregate stock valuation

Stores stock can be valued either at full value given by costing using one of the above costing methods, or by a 'written down' value to acknowledge the fact that some stock is no longer usable or saleable. There must be an amount of money deducted from the stock value to account for this. There are two ways of achieving this. The first is to depreciate the value of stock of each item. The longer an item is in stock the less it must be saleable. The rate at which depreciation is allocated will depend on the type of stock. Fashion goods can lose value very quickly. Aircraft spare parts depreciate very slowly. The simplest and most common way of calculating depreciation is to deduct a portion of the item value for each year it remains in stock. Taking 25% off the original value each year would mean that items have no value after 4 years and would call forth acknowledgement by the accountants that inventory control have made the expected mistakes and

Table 5.6 Alternative methods of determining unit stock value

Date (week)	Transaction Type	Quantity	Unit cost	Stock balance (quantity)	Cost of next item to be issued FIFO	LIFO	Standard cost	Average cost	Replacement value
	*	10	20	10	20	20	24	20	20
1	Demand	3		7	0	20	24	20	20
	Order	10	26						26
3	Demand	2		5	20	20	24	20	26
	Delivery	10	26	15	20	26	24	24.0	26
4	Demand	4		11	20	26	24	24.0	26
5	Demand	1		10	20	26	24	24.0	26
	Order	10	28						28
6	Demand	6		4	26	26	24	24	28
7	Demand	3		1	26	26	24	24	28
	Delivery	10	28	11	26	28	24	27.6	28
9	Demand	4		7	26	28	24	27.6	28
	Order	10	24						24
10	Demand	2		5	28	28	24	27.6	24

* Existing stock of 10 is valued at $20 each.
The standard cost set by the accountants is $24, but purchase costs are higher.
There will be a purchase variance of $70 when the outstanding order is received.

have squandered company profit on useless stock. It is not a healthy situation when inventory managers are forced to minimise the value of stock-holding because of poor control.

Writing the value of non-moving stock down is unfortunate but essential. The accounting code of practice requires stock to be valued at 'the lesser of its purchase cost and its realizable value'. This means that items bought at a discount have to be valued at a discount and items which have no likely use have to be valued at zero.

A second way of reducing the value of stock is to allow a Provision for the total value likely to be lost in a year. This provision can be based on a detailed depreciation calculation or a simple assessment. Whereas depreciation is applied to individual items, provisions are applied at a global level. This enables the provisions 'reserve' to be used for whatever accidents happen to cause stock obsolescence (sudden changes in legislation, contract loss, etc.).

Professional stock controllers do not use write down as a method of reducing stock value because this only transfers funds from company profits to pay for the write down. (The proper methods for reducing inventory value were discussed in chapter 3 in the section on *Practical methods of reducing stock*.) For stock control purposes it is therefore good inventory management practice to value stock at its full value and to apply the depreciation reserve or provisions later for the financial accounts. This ensures that the slow moving, high value stock receive some priority in control activities.

Skills and systems

Systems should not only be designed to perform a task, but also to enable the user to achieve the objectives most conveniently. The term 'user friendly' is often applied but seldom true, and the usability of a system should be geared to suit the type of person expected to use it and the knowledge, skill and time they are likely to devote to it. Where a complex system is in use, all operatives must be fully trained to understand it, otherwise they will not use it efficiently.

The more professional the user, the more complex the computer system can be and the better the results achieved by its operatives. This is true from the user options point of view. A routine operator may only require an uncluttered transaction screen with the system set up for fast input or enquiry, showing the relevant information and nothing else. However, if information is to be accessed by less frequent users, they will still need to have a thorough understanding of what the system can be made to do. If they don't understand it, they won't be able to use it effectively. Further, although systems should be straightforward and user friendly, their underlying logic should be complex enough to cope with the many variables which can occur. For instance, a user with a seasonal demand pattern will have to disregard a system only providing a simple minimum stock level to help in reordering. What they require is a seasonal forecasting system using percentage weighting, base series or Fourier analysis. Just as it is not necessary to understand the circuit for electronic ignition if you want to drive a car, so it is not necessary for the user to understand underlying computer logic as long as the method can be applied, the answer is good and the forecast type is easy to switch on and off.

Two distinct features for choosing a successful system are apparent simplicity geared to user skill and system sophistication geared to the importance of the application. Successful implementation develops the right balance between systems, procedures and the user's knowledge of a particular application. As an example, take a computerised parts list. A customer requires an item but does not know the part number. The old hands in stock control with many years experience will just quote the item number and look at the stock record because they know it. A new starter requires support in finding the number. How can this be achieved? There are basically three methods, namely training, support procedures and system features.

Training

If the user has an understanding of the application and memory of the codes, the system will work effectively. Training is essential for providing the understanding and knowledge. Inventory personnel need specific job skills which enable them to gain the full benefits of their systems and contribute to operating effectiveness. Investment in training is minuscule compared with the direct inventory savings.

Support procedures

Where the user of the system is unlikely to have knowledge of the application (e.g. technical knowledge by sales clerks), the company can provide the necessary information through supporting procedures. These can be paperwork or other aids which enable the user to follow a logical sequence with which to provide or gain the necessary information.

An example of this could be a person who often receives enquiries from customers needing advice on which stock item to buy. This person has either to gain sufficient technical knowledge of the stock to give advice, or to pass the item on to an expert, or to have the information provided for them in an appropriate form. A simple method of supporting this person is to provide a 'decision tree' for routine queries consisting of a set of questions. The answer to one question directs the user to the next question until the query is resolved. A person without detailed knowledge of a subject can still be effective using such support procedures.

Development of these procedures has to be overseen by a highly skilled person. Thus there is an initial cost, but the procedure can then run and advice can be given without the same level of expertise. This is a cost saving where the situation occurs regularly.

System features

Stock management systems can be considered as either stock recording systems or stock control systems. Stock recording systems concentrate on physical movements, recording issues, receipts and orders. Many of these systems also value the stock and keep a transaction history. 'Maximum' and 'minimum' levels can be input by the inventory controller and these levels and the order quantities are left to the discretion of the user.

Stock control systems take this data and provide decision support for inventory managers. Stock can be analysed into ABC classes and the historical information used to calculate the (weighted) average demand and the variability. Once policy on customer service and supply lead times has been decided, the system calculates the safety stocks and order quantities. The outputs of a stock control system are:

- Purchase schedules and orders.
- Expediting notes for suppliers.
- Management control information.
- Exceptions – unfulfilled customer orders.
  - Forecast predicting the wrong demand level.
  - Wrong supplier lead times.
  - Excess and obsolete stock analyses.
  - Options for special investigation identified by the user.

A successful system will produce information which can be used without amendment by the inventory controller. Some data may need modifying because the full information was not provided to the system, but this should be rare if the system is operated properly. In practice, most systems fail to provide this full specification, resulting in time being wasted on routine analysis and interpretation. This time could be much better employed on customer and supplier liaison and ensuring that the system parameters are accurate. It is therefore important to use a system with stock control features and to use these features to manage the majority of inventory.

6

Safety stocks

Learning from history

Collecting historical information

Stock is held either because it is convenient to buy in bulk or because the item is required faster than the supply can provide it. In the latter case there is some uncertainty as to the quantity required and some safety stock is needed. In chapter 2 the balance between the conflicting requirements of good service, low inventory costs and small operating costs were discussed. Now the means of achieving the controls can be examined.

The amount of safety stock held in an organisation depends upon three main factors:

- The variability of demand
- The reliability of supply
- The dependability of transport.

The general approach to this situation is to set stock levels to cover the normal variability of demand and to adjust the provisioning variables so that they are relatively insignificant. Quality initiatives have improved the supply situation but it is usually found that the major uncertainty is caused by customers and their unpredictable requirements.

Store records show the movement of stock in and out of the warehouse and such historical information is essential for evaluating what level of stock to hold (unless the customer provides firm orders). However, the best guide to the appropriate stockholding is the amount of demand, rather than the number of issues.

Individual stock movements logged on the stock records have to be analysed into movements per time period, normally weekly, so that weekly usage statistics are produced for each stock item. In cases where the demand is small or variable, then monthly periods (time buckets) can be

Table 6.1 Demand history

Period	1	2	3	4	5	6	7	8	9	10	11	12	13	14	15
Demand	23	35	28	19	34	25	41	15	39	33	28	48	31	38	28

When using demand history in forecasting, it is conventional to neglect the demand information from the current, incomplete period unless the demand is extremely large.

used. For fast moving and rapidly varying demand (such as chart music), daily time buckets are required.

The first step in analysing demand history is to collect the total demand for each period, as shown in Table 6.1. Sometimes such a table is used to estimate the stock levels and order requirements, whereas a little more understanding would enable accurate stock levels to be established.

The demand pattern can be displayed on a frequency distribution graph or histogram which shows for how many periods the demand was at a certain rate. Figure 6.1 shows that the demand was between 25 and 30 at three times in Table 6.1 and that it was between 40 and 45 once. This histogram gives a good idea of the usual demand rate and the spread of rates. In Fig. 6.1 rate intervals of five per period were chosen (11–15, 16–20, 21–25, etc.) and the figures on the horizontal axis represent the maximum figures in the range). As can be seen the most common demand rate is 31–35.

If the demand comes from a similar type of customer, the variations in orders will generally follow the same characteristics, although they are unpredictable. These demands form part of a population. If this assumption is correct then the properties of this whole population can be used to predict the demand pattern. The most common demand pattern is the Gaussian or normal distribution shown in Fig. 6.2. This is an idealised form of Fig. 6.1 for a large number of periods of demand. Using the small amount of information given in a table like Table 6.1, the shape of the curve in Fig. 6.2 can be calculated quite accurately.

Demand patterns can be assumed to be normal. (Even if they are not, taking them to be normal is much better than simpler methods of control.) As can be seen from the distribution graphs, most of the time the demand is near to the average. Sometimes, but less often, it is much more or much less. The vertical scale of the graph is frequency – how often that value occurs. The horizontal scale is really measured from the middle (the mean) and shows the values at which the populations occur. In practice, careful batching of the data is needed to produce a distribution like the one shown. In general, there is only a small amount of data to work with and this is used to fit a normal distribution.

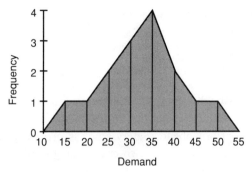

Figure 6.1 Demand distribution.

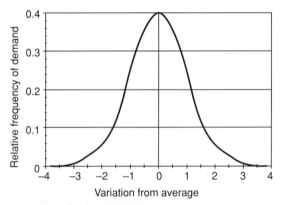

Figure 6.2 Normal distribution.

A key property of the normal distribution is that it can be described using only two parameters, the average and the average width. Calculation of the average value is relatively simple. Measuring the width is more difficult.

It is of no use to take the highest and lowest figures, as these are usually the result of atypical situations (stockouts, one-off demands, holidays, etc.) It is better to take values which are, say, a quarter of the way in from the extremes of the sample (quartiles) or apply a more rigorous approach which uses all the available data. In general the measure of width used is standard deviation, although in stock control a simpler approach is used, namely mean absolute deviation (MAD) (see section on *Standard deviation* below).

Of course more complicated demand patterns do arise. Occasionally there is more than one normal distribution pattern mixed in the demand for an item. For example, a product could be sold in a local market in single units, whereas for export it is boxed in 50s. The demand pattern will then consist of two normal distributions, one with small fluctuations in the demand pattern and the other with a more sporadic pattern.

Reasons for safety stock

The normal amount of stockholding can be determined by statistical methods which rely on history to predict the future and assume no change in circumstances during periods ahead. Safety stock is primarily to cover random variations in demand, but it can also cover many other situations such as:

- Supply failure
- Production shortfall
- Transport failure
- Slow, unreliable or inaccurate information
- Any other source of disruption of service.

Safety stock is the buffer between supply and demand. It decouples customer service from manufacturing and enables each to operate independently and more effectively.

Normal demand patterns

Stock availability

There is a need first to identify the major demand and supply variables discussed ealier in this chapter and to quantify their variability. The need to cover for changeability of demand is a central task of Stock Control. Unexpected demand requires a level of safety stock to cover for it. In fact using the normal distribution, the unexpected demand can be anticipated, or at least taken into account.

The more safety stock the better the customer service. Going back to the sample data in Table 6.1, a stock of 100 would probably be far too many and a stock of 35 too few. With a stock of 45 it looks as though there will only be a shortage once in 15 periods. But a surer way of finding this out would be to use all 15 sets of data to give an accurate normal distribution and then to find the risk of running out of stock.

Consider a supplier who arrives at the start of each period and tops up the stock to the required amount. One of the items supplied has the usage history shown in Table 6.1 and a unit cost of £1000. If the average usage amount, 31, is held in stock, then the expectation would be that the stock

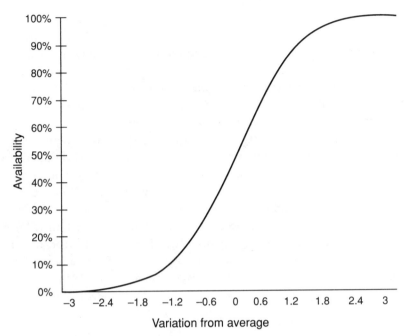

Figure 6.3 Availability of safety stock.

would run out in half the time, since the demand exceeds the average half the time. (This happens seven out of the 15 times in the sample data). If there were no stock at all then the stock would run out every time and of course if there were an infinite number in stock then there would be 100% availability. As the stock increases, the availability increases, and this relationship is shown in Fig. 6.3. As safety stock is added above the average demand, the availability increases rapidly, but once the availability becomes high, it requires a lot of stock to improve the situation.

Standard deviation

There are several ways of finding the typical variability, but there are two measurements which are generally used, namely standard deviation and mean absolute deviation (MAD). These are different ways of measuring the same thing, just as wealth can be measured in dollars or pounds, but the total value is the same. There are relative merits for both measurements. Standard deviation is theoretically the correct measurement, but people find it more difficult to understand and work with. Standard deviations should be built into computer systems. MADs are much more user friendly. They are easy to understand and arguably easier to bias. (The reason for wanting to bias them will be discussed in chapter 10.) MAD is a good enough measure in practice. The standard deviation will be described first.

The basic premise for safety stock is that the differences between the forecast and the actual demand are random. These random variations in demand usually follow a 'normal' distribution (see Fig. 6.2). This is symmetrical about the average value. The size of the variations differs from one stock item to another and consequently the width of the normal distribution differs also. This width, measured by the standard deviation (SD) is calculated from the sum of the squares of x, where x is the error, the difference between the actual value and the average (or the variations of the individual demand figures from the average). Mathematically this is:

$$SD = \sqrt{\left(\frac{1}{N} \times \sum x_i^2\right)}$$

The SD is calculated by adding up the squares of the errors from a sample of N values, in our case periods of history. Using the history given in Table 6.1, this gives an SD value of 8.4. If there is only a little history, then N is low and the value of the SD will be relatively inaccurate. This inaccuracy (the standard error of the SD) is given by a standard error of $SD = \sqrt{SD}$.

Mean absolute deviation

The MAD is a simple assessment of the variability of the demand pattern. Like SD, MAD is the basis from which safety stock is calculated. It is easiest to explain using an example.

For a demand during successive periods of 5, 14, 6, 15 the average demand is 10. The swings that occur about that average are −5, +4, −4, +5. Adding up these values gives 0, which shows that 10 is the true average. To measure the variability, it is the size of the errors not the signs which matter, so forget the signs and add up the numbers. This gives 18 in total, or an average of 4.5 for the four periods and is called adding the 'absolute deviations' or 'moduli'.

A second demand pattern has the form: 9, 12, 8, 11 for the four periods of history, and an average of 10 as in the first example. By the same analysis, the absolute deviations are now 1, 2, 2, 1 and the MAD is 1.5 (6/4). Obviously the safety stock required in this case will be considerably smaller than in the previous example, in the ratio 4.5 to 1.5.

The MAD is calculated by this method, or by taking the last value of MAD and updating it with the latest absolute deviation. The formula for MAD is:

$$MAD = \frac{\text{sum of absolute deviations from mean}}{\text{number of periods included in the sum}}$$

For a normal distribution, MAD = 0.8 SD. The MAD is easily calculated but to avoid the nuisance of recalculating each time and to give a better value, an exponentially smoothed MAD, is normally used (see chapter 10, section on *Weighted averages*).

Choice of deviation measuring technique

With any statistical distribution, such as the difference between forecast and actual demand, there is a variety of standard ways for assessing the dispersion, such as quartiles, MAD and SD. Approaches using the greatest and least demands are unreliable since the extreme demands usually result from an abnormal occurrence. Basing the variability (and safety stock) on a stock cover calculation is a common but incorrect assessment. This assumption gives completely the wrong stock balance.

There is a more serious debate about the relative merits of MAD and standard deviation. MAD is easy to understand, to calculate and to set up on a computer. SD is theoretically more correct, is used throughout statistics and is available as a preprogrammed function on spreadsheets. It is more difficult to calculate and understand.

For improved safety stock evaluation, the assessment of variability should be biased towards using the most recent data and the value should be exponentially weighted (see chapter 10, section on *Improved values for MAD*). This can be done very effectively with MAD whereas the SD has to be converted to the variance (= SD^2) in order to use the weighting technique. Exponentially weighted MADs are therefore preferable to SD. Their use results in improved safety stocks and offers a better selection of forecasting techniques to meet future demands (see *Choosing the best forecast* in chapter 10).

The smoothing factor which is most commonly used for MAD is 0.1, since there is normally no rapid change in the demand variability. In practice this may prove to be a little insensitive and 0.15 is often found to work better.

Table 6.2 Summary of deviation measurement techniques

Historical evaluation technique	Comment	Relative quality rating
Stock cover	Only for the amateur A major improvement on continuous inspection	1
MAD	Simple to understand and calculate Good for most situations	10
Standard deviation	Theoretically more accurate Available on spreadsheets	13
Smoothed MAD	Gives better results Also suitable for seasonal demand Easy to put into spreadsheets and software	25
Smoothed variance	Should give best values	30

For seasonal demands, the variations in size of the demands are likely to be greater at the maximum and smaller at the minimum. MAD can be considered as a percentage of the demand level. Using this 'percentage MAD' approach enables an appropriate and up-to-date variability to be maintained and provide the appropriate safety stock levels. The use of exponentially weighted variances should give better values in theory, but is seldom used in practice. Table 6.2 shows a simple summary of the options. There is an arbitrary quality scale included to illustrate the relative merits of the techniques.

For appropriate safety stocks, interpret the historical data, measure demand not sales and respond to changing variability in demand.

Evaluation of safety stocks

Risk measurement

The reason for having to use SDs and MADs is that they are the only sensible ways to work out stock levels. The shape of the normal distribution curve measured in terms of standard deviations (or MADs) is always the same. This is the shape of the curves shown in Fig. 6.2 and 6.3. These graphs indicate that when stocks are held at average usage, 50% service is given. The other values behind Fig. 6.3 are given by the customer service factors shown in Table 6.3. The customer service factors are shown in the table for both SD and MAD. Either measure is acceptable and will give similar answers.

Table 6.3 Customer service factors (assuming a normal distribution)

Desired service level (% periods without stockout)	Multiply SD by	Multiply MAD by
50.0	0.00	0.00
75.0	0.67	0.84
79.0	0.80	1.00
80.0	0.84	1.05
84.13	1.00	1.25
85.0	1.04	1.30
89.44	1.25	1.56
90.0	1.28	1.60
93.32	1.50	1.88
94.0	1.56	1.95
94.52	1.60	2.00
95.0	1.65	2.06
96.0	1.75	2.19
97.0	1.88	2.35
97.72	2.00	2.50
98.0	2.05	2.56
99.0	2.33	2.91
99.18	2.40	3.00
99.5	2.57	3.20
99.7	2.75	3.44
99.86	3.00	3.75
99.9	3.09	3.85
99.93	3.20	4.00
99.99	4.00	5.00

Adding one MAD of safety stock to the average requirement means that customer service increases from 50% to 79%, or when stocks are increased by one standard deviation, the service increases to 84%, leaving a 16% chance of stockout. By increasing the buffer stock to twice the standard deviation, service is increased to 97%. Additional stockholding increases the service towards 100% but the cost in terms of inventory level is high. Addition of three amounts of SD gives 99.86% and SD gives 99.99%, but these are increases of only 2.1% and 0.13%, respectively for the same investment in stock that gave 34% for the first standard deviation and 13.6% for the second. This raises the question of which service will be too expensive to give and whether the customer can tell the difference between 99.86% and 99.99% availability.

The availability level is measured by how often the designed stock level will meet all the demands during the period. Other measures would be how many items were short during a year, or how many customer orders were not completed.

Table 6.4 Demand pattern

Period	Demand	Stock	Purchase orders[a]	Demand variability
0		10		
1	3	7 →	10	
2	0	7	10	2
3	2	2 ←	10	
4	4	8		1
5	1	3 →	10	1
6	6	0[b]	10	3
7	3	0[b] ←	10	
8	0	10		2
9	4	6 →	10	1
10	2	4	10	

[a] The number 10 in this column indicates an order for 10 is placed but outstanding.
[b] Unable to meet demand.

An illustration

Taking an example, Table 6.4 shows a conventional stores movement record for one item, with receipts, despatches and supply orders for a stores. The average usage per period is 2.5. If the policy is to purchase sufficient stock to cover four periods, then the order quantity will be 10. If it takes three periods' lead time for the stock to be replenished, then replenishment orders need to be placed when the stock falls below 7.5 items. This is called the 'review level' because when the stock falls to this level the decision should be taken on whether to order. All too frequently an order is raised at some arbitrary 'minimum stock level' which bears no relationship to the current usage rate. As a result of assessing the demand, a purchaser will change the review level, place an order, or both.

The stock levels resulting from this are shown in the 'stock' column in Table 6.4. The orders are raised each time the stock falls below 7.5. (This review level is taken as a constant over the short time shown). Notice that there is a shortage in periods 6 and 7 resulting from the large number of sales in period 6 which coincided with the stock being below the stock review level. It is prudent to hold safety stocks to cover for these occurrences.

The amount of stock can be assessed from a statistical analysis of the demand pattern as discussed in the previous sections. The total sum of the variations from the mean is 15 for the 10 periods. This gives a MAD of 1.5. If the safety stock is now increased by 2 from 7.5 to 9.5, this should give added protection of 2/1.5 MAD or 1.33 MAD. This corresponds to an 86% safety stock. The stock review level is also increased from 7.5 to

Table 6.5 Demand pattern including safety stock

Period	Demand	Stock	Orders
0		10	
1	3	7 →	10
2	0	7	10
3	2	5 ←	10
4	4	11	
5	1	10 →	10
6	6	4	10
7	3	1 ←	10
8	0	11	
9	4	7 →	10
10	2	5	10

9.5 and the orders have been placed immediately at this level. The new pattern is shown in Table 6.5. The extra two stock has reduced the number of depleted weeks from two down to nil. It also meant that it was possible to supply all of the 25 items needed. With no safety stock it was possible to issue only 19. This calculation will be refined below.

Of course, from the statistics, this level of safety stock will cause a shortage in 14% of the order periods and this may not be good enough. Additional safety stock can be added to give the service level required. The higher the safety stock, the more costly is the stockholding, not only because money is invested in the stock, but also because an investment in stock may prove fruitless if the item is superseded or becomes obsolete.

Additional safety stock

Safety stock is normally calculated on the basis of unknown demand. The quality of supply from the manufacturers may also prompt an increase in safety stock, either because the lead time for an item may vary between placing one order and the next, necessitating extra stockholding or because the supplier's delivery performance against the lead time is unreliable and delivery is more often late than early. An allowance in stock has to be made for this factor. If stock has to be held against poor supply, then theoretically a supply performance MAD could be calculated and a safety stock worked out. This supply safety stock would not be additional to the demand safety stock, since the same safety stock would cater for either high demand or late delivery, but not both. It is found that the addition of MAD from different causes can be added using Pythagoras' theorem (in a similar way to adding variances).

In dealing with the supplier, the supply pattern should be a result of compromise for both organisations. Whoever is the cause of the extra safety stock should incur the cost of it. Minimum stock is held when each party can trust and accept the information provided by the order.

Slow moving items

For a large number of stock items the usage rate is low. This includes obsolescent items, non-standard lines and spares. For these the 'normal' distribution is squashed against zero usage and is replaced by a skewed distribution. The chance of an item being required is governed by Poisson's distribution law. Compared with the normal distribution, the Poisson distribution has a longer tail where there is an enhanced (but small) chance of high demand during a period. To cover for this, a higher stock would be required where high levels of availability are needed.

An important feature of the Poisson distribution is that

$$SD = \sqrt{\text{average demand}}$$

This relationship simplifies the review level formula for slow moving items. When using the Poisson approximation, the usage events have to be independent (i.e. issues in ones to unconnected demand). If this is not the case, then SD of sales orders equals the square root of average number of sales orders.

Using the calculations developed in this chapter and applying the lead times discussed in the next chapter provides the basic method for inventory management for the majority of stock.

Excess and obsolete items

In practice, stocks are never balanced ideally and some items become excessively slow moving. These items can be described as 'excess' or 'obsolete'. One of the areas which generates little interest and lots of cost is obsolescent items. The definition of obsolete should be that there is no use for the item. This information is normally only available when the customer changes to a direct and improved substitute. In many instances, goods are classed as obsolete as a result of a long period with no demand, so the practical definition is usually an item is obsolete if it has not moved for 'x' months.

The selection of x varies with the type of item and market, and the position in the supply chain. It is relatively easy in fast moving consumer goods to get rid of items which have not moved for 3 months and this may well be the definition in this environment. Suppliers of heavy machinery and spare parts, where world demand has been zero for 2 years, may not regard demand to have finished and they will still desire to keep stock (only one!). In this case failure to supply quickly may have important strategic or economic effects and supply lead times are long. A mechanical part on a machine which fails every 7 years or so is not obsolete, but has a very low usage rate. The decision to purchase such slow moving stocks is one problem, the decision to throw away existing stock is entirely another. Each company has to decide on a proper obsolescence policy for disposing of non-moving items.

The 'not moved since' analysis can cope with non-movers, but there are other useless items, namely the excess items. If the stock of an item is 1000 and the usage is one per month, the item is certainly not obsolete, but may be before the stock is used up. Therefore the definition of excess could be: excess items have stock of more than 'y' months usage. Again the value of y depends on product, market, position in the supply chain and also on the Pareto class of the item, since the stockturn definition is a limit on the stock investment.

A diagram with cut off values x and y (Fig. 6.4) shows simple control of excess and obsolete stock. It allows accountants to identify how much money to put aside to cover likely stock disposal. Writing down

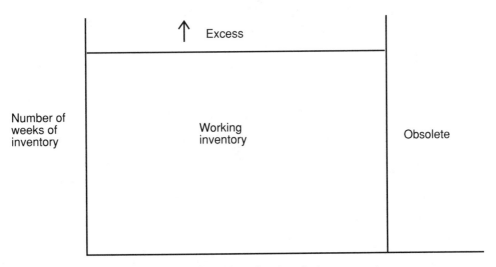

Figure 6.4 Definition of excess and obsolete stock.

inventory value is, of course, an admission by the stock controllers that they failed!

In theory the values of x and y could be different for each item. For example, a stock of helicopter rotor assemblies could be held in stock a long time before use, but there is a mandatory requirement to change and repair these units, so the stock is not obsolete, just slow moving. Whether it is excess depends on whether the items are now units (bought too early) or repaired units (which would be in inventory anyway). In practice, therefore, the cut-offs shown in Fig. 6.4 are used and some allowance is made for items with stock of one.

The first step in dealing with the phase-out of stock lines is to identify early the lines in this terminal condition. These lines can then be included in the ABC classification as 'O' or another suitable flag to inhibit normal ordering. The disposal of excess and obsolete stock has to be carried out carefully and professionally to avoid a large loss in profits. Before selling a company, the value of obsolete stock items has to be reduced since it is illegal to overvalue the company and the stock can be a major part of this value.

7

Setting the right stock levels

Simple assessment of review levels

The basic calculation

Our aim in stock control is to maintain a balanced inventory so that customer service for each stock item is maintained within its proper limits. The time at which we can influence the stock levels most effectively is when we order and it is at this point that the major opportunity occurs for ensuring a balanced inventory. This is the major control mechanism for ensuring a balanced inventory. It is also the point at which customer satisfaction or excess inventories are created. The technique for calculating the time for ordering inventory is crucial to balancing the conflicting pressures.

Originally stock recording systems used to include stock control levels as 'minimum' and 'maximum' stock levels. These basic concepts are still applied today as a simplistic approach where tight control over stock is not necessary. The use of a minimum stock level for order control is not sensible, since the minimum occurs immediately before delivery. Items have to be ordered well in advance of this, so control is through a 'reorder level' not a minimum stock. However a minimum stock level is vital to ensure that there is warning of low stocks. In a stock control system, therefore, there is a need for both reorder level and minimum stock level. These are called the 'review level' and 'safety stock'.

The review level is triggered by the information from the system that stock is low (i.e. it has fallen below the reorder level) and makes a decision as to whether to order. Where the demand has increased or lead times increased, orders are necessary; where the lead time is reduced or demand has diminished, then there is no purchase. Since the likelihood of each is equal, the trigger level is treated in practice as a review level.

Minimum stock has also been replaced by the safety stock, which is calculated as discussed in chapter 6, and allowing for lead time (the analysis

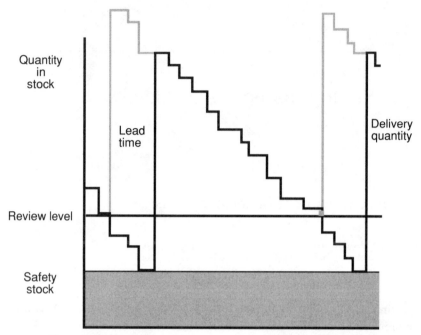

Figure 7.1 Stock controls.

is shown below). Maximum stock is a result of negotiation with suppliers on delivery quantities. The maximum stock level should be a result, rather than a direct control. It is therefore necessary to control the delivery quantities directly by using three parameters to carry out simple and effective stock control, as shown in Fig. 7.1:

- Review level
- Safety stock
- Delivery quantity

The starting point is the calculation of the safety stocks. A quick look at the historical demand for the two items illustrated in Fig. 7.2 shows that the second will require a great deal more safety stock than the first if the customers are going to be equally satisfied. The average demand rate is the same in each case. The more the variability of demand, the more safety stock is required. Therefore the safety stock does not depend on the demand rate. (This means that safety stock calculated as number of weeks' demand is simply wrong.) The safety stock for periodic delivery items is given by multiplying the customer service factor by the standard deviation of the demand during the period.

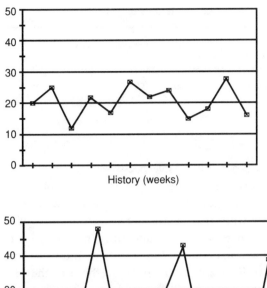

Figure 7.2 Historical demand patterns.

If the demand is 30 per month and the supply lead time is 2 weeks then if there are not 15 in stock or on order, there will be a shortage. This is because the time to order is given by:

review level = usage rate × supply lead time

and to be safe:

$$review\ level = \left(usage\ rate \times supply\ lead\ time\right) + safety\ stock$$

The safety stock calculation discussed in chapter 6 showed that the stock depends on the variability of demand (MAD or standard deviation, SD) and the availability level required (customer service factor). Although the effect of supply lead time was not discussed, the safety stock also depends on the supply lead time.

Table 7.1 Variation of safety stock with lead time

	per day	per week	per month	per year
SD scaled proportional to lead time	0.4	2	8	100
SD scaled as the square root of lead time	0.9	2	4	14

Example

For an item with a demand of five per week ±2 (i.e. MAD = 2) what is likely to be the variation in demand per day, per month or per year? Two possible answers are suggested in Table 7.1. The average demand per period increases linearly with time period. The uncertainty of the demand, however, does not increase at the same rate. The average is more likely to be achieved over a long period of time rather than over a short period. The annual sales of an item are often similar to those in the year before, but from day to day the demand will vary widely. The weekly sales will also be quite widely dispersed but the monthly results are quite consistent. If the standard deviation were to increase in proportion to the time period (see Table 7.1), the inaccuracy in the long term forecast would be great. However because of the 'compensating errors' over longer time periods, the deviations are relatively small. In fact it is found (and can probably be proved) that the SD only increases as the square root of the time period. This is shown as the last line in Table 7.1.

This same effect can be illustrated in Fig. 7.2B. A short sample could give a wide variety of answers to the average use, whereas a longer sample will give a more reliable answer. In other words, the longer the sample period (in the case under discussion – the lead time) the less the relative safety stock need be. From chapter 6 this would be a square root relationship. As the variation in demand is largely a statistical effect, the square root relationship occurs frequently in stock control. The full formula for safety stock is therefore:

$$\text{safety stock} = \text{customer service factor} \times \text{SD} \times \sqrt{\text{lead time}}$$

For examples of customer service factor, see Table 6.3. This equation can then be used in the review level calculation above.

Example

The full calculation of review level can be carried out for the item with the demand history shown at the start of chapter 6 (Table 6.1). The average was 31 and the MAD was 6.8. Some practical information is now required, namely, the supplier's lead time and the availability level to be offered to customers. Assuming that the lead time is three periods and the required customer service is 90%, the customer service factor for 90% customer service is 1.6 MAD from Table 6.3. First the safety stock has to be calculated:

$$\text{safety stock} = \text{customer service factor} \times \text{MAD} \times \sqrt{\text{lead time}}$$
$$= 1.6 \times 6.8 \times \sqrt{3}$$
$$= 18.84$$

From this the review level can be calculated:

$$\text{review level} = \left(\text{usage rate} \times \text{lead time}\right) + \text{safety stock}$$
$$= \left(31 \times 3\right) + 18.84$$
$$= 111.84$$

This means that if the stock falls to 111, then more need ordering (assuming there are not enough on order already to cover the 111). If it is decided that 90% service is not good enough and 98% is more appropriate, then the customer service factor becomes 2.56, the safety stock 30.15 and the review level 123.15.

In reality, all the items do not need to be in stock to provide availability for customers as long as they are on order and due for delivery before the current stock runs out. In the illustration above, if the stock is currently 60 then there is a problem. However if there are already 100 on order, there is no need to raise another order. Therefore the condition for ordering is when:

$$\text{current stock} + \text{supply orders outstanding} < \text{review level}$$

To be exact, the supply orders which count here are only those which are due for delivery within the supply lead time. The purpose of the review level is to trigger replenishment orders at the right time. Of course triggering the order at the right time does not necessarily ensure that there is stock in the stores. The review level calculation described here is the most important part of stock control theory.

When to expedite supplies

From the purchasing point of view the supplier has to deliver on time and should be chased if the delivery is late. From a stock control point of view, we rely on the supplier delivering at a fixed lead time. However, if the supply is late, it does not necessarily cause a problem if the rate of issue has not been as fast as expected (and the stock is still relatively high). In other circumstances the demand may be higher than average and it would be convenient to receive delivery early. In the example discussed before, a 90% availability means that once in every 10 time periods a stockout is to be expected. When this happens it would be useful to try to enhance availability by rushing in supplies. Therefore, a warning is needed when the physical stock is getting low, not when the supply is late. As the supply should, on average, arrive when the stock gets down to the safety stock level (see Fig. 7.1), then the safety stock level is a suitable trigger level to remind the supplier that goods should be arriving. Sometimes this reminder will be after the proper lead time, if delivery is late, or it could be before the item is due when the demand is above average. In either case the safety stock is still available to use before there is a crisis. The condition for chasing supplier is therefore when the current stock is less than the safety stock. This should be a standard routine printout from the system used for automatic expediting.

Managing lead times

Lead time

Lead time is the time between a shortage occurring and items being available to maintain supply. The lead time for supply is a balance between what the supplier wants and what the customer wants. If the suppliers are powerful, they will specify a long lead time (such as a one year firm programme). If the customers have the upper hand, they will expect to be supplied immediately, without any commitment. Both of these extreme cases increase the total operating costs. The compromise is a lead time which gives the suppliers sufficient supply time, but does not force the customers to commit themselves further than they can predict. The accuracy of the forecast of their requirements reduces rapidly the further ahead they look. This produces an order pattern which is irregular and subject to change. Stocks are needed where the sales lead time is shorter than the supply lead time. If this can be altered so that supply lead time becomes less than sales lead time, stockholding can be avoided.

Lead times for purchasing are governed by production and transport, lead times for manufacturing result from capacity, material availability and production planning, and lead times required by customers depend upon availability and policy.

Components of lead time

Lead time can be analysed into several components, some of which are essential and some avoidable. The supply lead time comprises:

- Order review time – the intervals at which the low stock situation is reviewed (e.g. printout every day, hourly, weekly, etc.).
- Order processing time (purchasing and communication) – the time it takes to:
  - review the order
  - decide to buy
  - transmit to the ordering system
  - raise an order
  - gain the appropriate authority
  - inform the supplier (post, fax, email).
- Supplier lead time (vendors, manufacturing, buying and despatch).
- Transport time – transfer of the item from supplier to receiving bay.
- Receiving time – time taken for goods inwards and updating the stores records.

From this analysis, the actual supply lead time contains some elements of administration in addition to the lead time quoted by the supplier.

In stock control equations, the supply lead time is assumed to be constant. It is very important that the estimate of lead time in the system reflects the true lead time. If the supply lead changes, then the lead time in the calculation should be changed to reflect this. It is a continuous task and part of the maintenance of the stock systems. Changes can normally be made by groups of items from the same source, or by using the actual lead time to update an (exponentially weighted) standard lead time.

Setting the lead times

If the lead time on the control system is set too short, this will lead to stock-outs, since the stock level will be kept too low. If the lead time is set over-cautiously (long), excess stock will result. The compromise for setting the lead time is a fine judgement for the inventory controller.

Minimising the lead time

Longer lead times make forecasting more difficult and increase the safety stock. Reducing the lead time is therefore desirable and can lead to stockless supply if the supply lead time is short enough. It is part of inventory management, therefore, to work on reducing the overall lead time.

One of the simplest ways of reducing the lead time is to specify it more precisely. 'Delivery in the month of June' as a request gives the supplier plenty of leeway. 'Delivery on 12th June at 11.18 a.m.' suggests an entirely different expectation and is likely to gain a much better delivery performance (during the day required rather than the month). Numerous experiments attempting this have shown that a majority of suppliers will improve their delivery precision greatly, simply by being asked.

Companies also find that supplier reliability is often affected by lead time. With a lead time of one day, if the supplier of an A class item is a day late, this can be a major issue. If the lead time is 12 weeks and the supplier is a day late, this is often expected, or even considered as on time. One day's worth of stock is held to offset this. The perception of delivery performance is as a proportion of lead time. However, from an inventory perspective, one day's worth of stock is the same value, whether it is added to an extra 6 weeks' worth or to one day's worth of total inventory.

From the analysis of lead time above, the total lead time should be minimised by:

- inputting information to stores and computer systems without delay.
- partnership with major suppliers.

It is important to minimise the time between:

- the stock getting low in the stores and the supplier receiving the purchase order.
- receiving the delivery and the event being displayed on the company computer system.

Once these elements of lead time are small enough, (see also chapter 8, section on *Purchasing procedures*) the suppliers' lead times should also be examined. Supply lead time consists of actual physical transport time, manufacturing time in some instances and organisational time. Lead time should be a negotiated time, not something that happens without manipulation.

Take an example. A supplier quotes a 6 week delivery. What does this mean? It could be:

- All the existing stock is allocated to someone else.
- There is a stockout because the safety stock was not high enough or the reordering procedure broke down.

- It takes 6 weeks to get these from their supplier.

In the first two cases, the 6 weeks are caused by priorities, first because the order is not jumping the queue and second because the supplier does not consider the order important enough to put in higher safety stock or improved systems. This is a matter for the supplier and for the customer to negotiate. The third case could be argued similarly. If their supplier is a manufacturer, then the 6 weeks could reflect the manufacturing time. However only a small part of manufacturing time is taken up with production. The manufacturing time per unit assessed by work measurement (time being worked on) is much smaller than the production lead time taken for the work to pass through the manufacturing plant. This is usually more than ten times the measured work content. There is, therefore, potential for manufacturing lead time to be collapsed to a much smaller timescale. However, this cannot be done on a one-off basis, it has to be done gradually for repetitive products because of its effect on capacity balance and replanning. Replanning manufacture has to be minimised in the short term or the plant becomes inefficient and delivery performance suffers. Again, collaboration with suppliers may enable them to reduce their own lead time.

Because of the large numbers of items in inventory, reduction of lead time has to be carried out for selected types of items. These are initially:

- A class items.
- Items where a small change in lead time can make them into supply-to-order items.

Effects of lead time

Lead time does not govern stockholding. Order quantity is not affected by lead time. It is possible to order 20 weeks' worth of an item with one week's lead time and one week's worth of an item with a 20 week lead time (as long as further orders are placed each week). True the review level is calculated using the lead time, but the review level determines only what order cover is required and not how much stock is held.

Lead time does affect the amount of safety stock to hold. The longer the lead time, the wider the absolute variation in demand and therefore the higher the safety stock for a given service level. Fortunately the amount of extra stock only increases as the square root of the lead time.

In summary, the effects of long lead time are generally:

- Forecasting demand to cover the supply period is more difficult.
- The reliability of the supplier may be poorer.

Errors in forecasting cause higher stocks to be held. If demand is certain over the lead time, then stocks can be very low. A short lead time means that:

- The forecast is more likely to be correct.
- Errors will be smaller.

Dealing with inconsistent lead times

With closer partnerships and higher expectations, inconsistent delivery is not the problem that it used to be. However customers are now more sensitive to supply reliability, so strategies should be considered for that supplier where delivery is inconsistent and changing the supplier is undesirable.

It is difficult to set reasonable stock levels or to provide good customer service if lead times are not known accurately. The policy on inconsistent lead times should be determined by the impact of the problem, and on whether the financial risk is high or low. For high turnover value items, (A class), the inconsistency has to be resolved through scheduling and working more closely with suppliers to smooth out the inconsistencies. The objective should be to smooth out the variations rather than to improve the updating, since each time the lead time is updated, the stock parameters will be changed and the order book modified.

For low turnover value items, (C class), the situation is not so important unless there are many inconsistent suppliers. The strategy for avoiding the problem for routine supplies is to use pessimistic lead time in calculations. This will cause higher stocks, which may be necessary if the item is important. However as the purchase quantity is probably large, there are only a few chances of running out and C class items may not be worth the investment in extra stock to provide the extra availability.

Effect of order frequency on safety stock

The safety stock calculated by the methods described earlier in this chapter are adequate for most practical situations. However the calculations can be improved because the risk of stockout is not constant all the time. Availability is good when there has just been a delivery. The risk of running out is greatest when the stock is lowest, i.e. while the goods are below the review level and a delivery is expected. The main risk during a year is, therefore, the risk during the lead times. This depends upon the order frequency. The more replenishment orders for an item that are placed per year, the higher the risk and the greater should be the safety stock. This factor is usually small compared with the benefits of ordering more frequently and reducing the inventory costs and risks of obsolescence. A simple addition to the calculation above suffices for control, as shown in the example that follows.

Example

Assume that the historical data for an item is as follows:

$$
\begin{aligned}
&\text{Annual demand} && = 10\,000 \\
&\text{MAD} && = 100 \text{ (weekly)} \\
&\text{Service level required} = 94\% \text{ (2 MADs) (measured as number of} \\
&&& \text{items short per year)} \\
&\text{Lead time} && = 4 \text{ weeks} \\
&\text{Order frequency} && = 6 \text{ weeks}
\end{aligned}
$$

So the total stock shortfall allowed = $1 - (94\%$ of $10\,000) = 600$ per year.

If there are $50 \div 6$ orders per year (8.3 orders per year) then the allowed stock deficit per stockout is:

$$
\frac{600}{50} \times 6 = 72 \text{ (assuming 50 sales weeks per year)}
$$

$$
\text{usage in the lead time} = \frac{10\,000}{50} \times 4 = 800
$$

The allowed stockout is therefore 72 in the lead time (LT) demand of 800. Therefore

$$
\text{service level required} = 1 - \frac{72}{800} = 91\%
$$

$91\% = 1.5$ (MADs) but this is for a 4 week MAD since the period is the lead time. Therefore

$$
\text{safety stock (SS)} = 1.5 \times 100 \times \sqrt{4} = 300
$$

and

$$
\text{review level} = 300 + 200 \times 4 = 1100
$$

The average stock level is

$$\text{SS} + 1/2 \text{ order quantity} = 300 + 1/2 \times 6/50 \times 10000$$
$$= 300 + 600 = 900$$

Now if the order frequency is changed to 10 weeks, then there are now five orders per year

$$\text{allowed stock deficit per stockout} = \frac{600}{5} = 120$$

The usage in the LT = 800 (as before), but because there are less risk periods.

$$\text{service level required} = 1 - \frac{120}{800} = 85\%$$

85% is 1.2 MADs (i.e. four week MADs). Therefore

$$\text{safety stock} = 1.2 \times 100 \times \sqrt{4} \text{ (in weekly MADs)}$$
$$\text{safety stock} = 240$$
$$\text{review level} = \text{SS} + \text{usage in LT} = 240 + 800 = 1040$$
$$\text{average stock level} = 240 + 1/2 \times 10/50 \times 10000 = 1240$$

The review level and safety stocks are less, but the average stock is more because of the larger order quantities. If the order frequency is changed to 4 weeks, the service level per stockout becomes 94%, as expected since there is continuous exposure to stockout risk. The safety stock becomes 400, the review level 1200 and the average stock 800.

Target stock levels

Application of target stock levels

The target stock level (TSL) is a maximum stock level which can be used to calculate the order quantities. It differs from the review level approach to ordering because review levels fix the order quantity and vary the order frequency and TSLs fix the order frequency and vary the order quantities. When ordering regularly from a supplier, the stock controller can have a cyclic work routine, (such as always ordering from supplier X on Thursdays) and TSL approach is therefore good to use for the regular demand items and is recommended for A class products.

The TSL is a level to which the stock is topped up (theoretically) when the cycle time for ordering arrives. The process for setting up the ordering procedure is first to establish a routine for placing orders with the various suppliers by specifying the day each week or month on which the order is going to be dealt with and second, on the allotted day, to review the stock of all the goods supplied by that supplier. The amount which is to be ordered is given by the formula:

order quantity = TSL − free stock − supply orders outstanding

Here the 'orders outstanding' are of course only those orders which are due within the current supply lead time. Orders booked further ahead are not included in the calculation.

An example is given here:

TSL as calculated = 34
current free stock = 16
outstanding order due tomorrow = 5

Therefore

order quantity = 34 − 16 − 5
= 13

Calculation of target stock levels

The calculation of the TSLs is similar to that for the review levels. An extra factor is included in the calculation of the TSL. The effective difference is that stock replenishment now needs to cover the period until the review cycle comes round again. If the review cycle were a month and the supply lead time only one day, then the TSL would have to cover one month's extra usage (plus safety stock) because the next order would be placed only in one month's time. Thus the formula for the TSL is given by:

$$\text{TSL} = \left[\text{usage rate} \times \left(\text{lead time} + \text{review period}\right)\right] + \text{safety stock}$$

and there is a similar adjustment to increase the safety stock level:

$$\text{safety stock} = \text{customer service factor} \times \text{MAD}$$
$$\times \sqrt{\left(\text{lead time} + \text{review period}\right)}$$

The demand 'in review period' is included because the stock level may not be checked continuously or deliveries may only be made once per month. If we do not include the extra time in these processes then the risk of running out of stock is increased. An example is given:

review period for an A class item = 1 week
supply lead time   = 5 weeks
average demand rate   = 4 per week

MAD   = 2.5 measured on a weekly basis

required customer service   = 90% (this gives a customer service factor of 1.6)

Therefore

$$\text{safety stock} = 1.6 \times 2.5 \times \sqrt{(5+1)}$$
$$= 9.8$$

and

$$\text{TSL} = 4 \times (5+1) + 9.8$$
$$= 33.8$$

Applying target stock levels

The TSLs are levels of stock to which the free stock never rises in a stable situation (see Fig. 7.3). The actual stock will rise to a maximum level which is:

$$\text{TSL} - (\text{usage rate} \times \text{lead time})$$

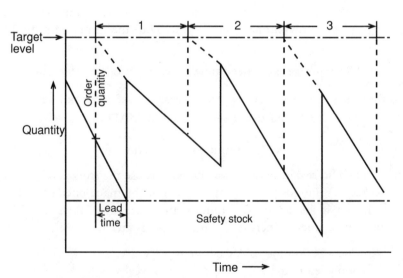

Figure 7.3 Target stock levels.

TSLs can be used for long review periods (e.g. one month) with short lead times, or of course for short review periods (1 week) and long supply lead times. In this latter case, there will always be one or more orders outstanding from the suppliers. This does not affect the validity of the calculation.

It is important to keep on adjusting TSLs to match the usage rates and lead times that are experienced or else this approach can lead to last minute purchases and extra work. It is important to ensure that the stock does not run out without warning during the order cycle, which is why it is essential to monitor continuously those items where the stock is falling below the safety stock level, calculated as in the section on *Target stock levels*.

8

The changing role of purchasing

Modern supply practice

Improving the supply

The development of supply chain management changes the relationships between companies providing goods for one another. This will affect the role of the purchasing function. Purchasing has traditionally been an order-by-order process based on obtaining the best combination of price, quality and delivery (availability). Estimates are requested, best options are negotiated, orders are placed and delivery is monitored.

This approach has been eroded over the years and the time consuming task of obtaining estimates is now not usual in routine material supply. The development of supplier relationships has introduced longer term arrangements involving contracts, call-offs and schedules. This has changed the role of the purchasing function. Instead of individual orders, supply is negotiated less frequently. This has meant a different style of purchasing activity because low level routine negotiation is being replaced by high level contracts. The number of suppliers is also reduced.

These changes have a major effect on the type of expertise required in purchasing. The traditional type of 'buying' is being superseded by a combination of high level contract management and supply scheduling as part of the material control activity. This means that materials management will absorb the purchasing operations as part of control and scheduling. Why is this now practical and possible? The answer lies in the current supply environment:

- Total quality
- Supplier partnerships
- Logistic control
- Linear production

- Single sourcing
- Supply chain management.

The new role of purchasing in a company is in negotiation – a skill which inventory management has been practising internally for a long time. Purchasing no longer requires the wheeler dealer, buying from a favoured source, hard hitting on price and always with a good idea of where to go for a cheap deal or a quick supply. Purchasers must now be contract managers, working in partnership and shrewd enough to balance the wellbeing of their company with the aspirations of their supply partners. They will have fewer suppliers and maintain multilevel collaboration within the supplying and user company. In fact in some companies the purchasing agreements will be carried out by top executives, as they were in the early days of business enterprise. This situation arises because of the improvement in communication between companies and the better organisation of the supply chain.

The purchasing environment

The environment in most businesses has changed from a supplier led to a customer led operation. This has resulted from the increase in international trading, multinational competition and improvements in transport systems.

The development of the customer led market has caused major changes in supply structures. There have been significant developments in:

- Supplier relations
- Quality management
- Stock control philosophy
- Supply chain structure
- Number and type of supplier
- Distribution
- Administration.

There has been a gradual change of power from supplier to customer leading to changes from a product-based supplier to a service-based supply (see Table 8.1).

This means that the requirements for suppliers and customers have changed (see Table 8.2):

Single sourcing

A major development in purchasing has been in the area of supplier relations, with a pressure to decrease the number of suppliers and to develop

Table 8.1 Changing attitudes to service

Service was supplier led	Service is customer driven
Manufacturer provides products	Customer requires solutions
Supplier provides narrow range of options	Customer sources wider range from each supplier
Non-standard individual items	Items are commodities, system components
Generic service	Different service for each customer
Delivery when available	Fast reliable service
Provide products	Packaged items ready for customers to use or sell

Table 8.2 Effects of changing attitudes to service

	Issues for service providers	Issues for users
Supplier led Service	Supply lead times Single logistic chain High demand rates Good margins	Variety of suppliers Many orders to control
Customer led Service	Good availability Wider type of demand Stock and supply risks Commodity type supply	Convenience Good control and price Single source Supplier partnerships

partnerships with them. This stems from the need for improved supply and has led to tighter control and higher quality standards. One solution to the supply problem is seen to be a single supplier for each item, providing the supply risks can be minimised.

The approach toward single sourcing can be made in the same way as any other quality improvement process:

• Improve process until a problem is encountered.
• Identify the cause of the problem.
• Find means of solving or avoiding the issue.
• Advance to the next problem.

Managers often consider that their own situation is not appropriate for single source supply. Traditionally, companies purchased items from a variety of suppliers, choosing a supplier at any time on the grounds of cost or availability, or less often, quality. Such a formal arm's length relationship, with the customer giving an 'order' to the most fitting supplier, was more costly and time consuming than a single sourcing arrangement and did not necessarily give the customer the best service. A supplier who is not confident of the customer will:

- Present a price from which to negotiate
- Only meet the customer's specification and no more
- Offer improvements only under duress
- Provide no add-on services
- Consider only the short term profit
- Fear losing the next order.

This negative attitude stops the supply chain from being coordinated properly. The exception to this situation is where the customer is so dominant that the supplier is treated like a part of the customer's company and is dictated to by the customer. This is the situation in many Japanese companies. The normal objections to single sourcing are:

- Supplier has too much power
- Inflated prices
- Risk of poor delivery performance
- Lack of flexibility
- Supplier may cease trading
- Divulging sensitive information
- Risk of quality problems.

These are real fears for those not accustomed to single sourcing and present the first barriers to be crossed.

There must be compensating benefits to encourage companies to adopt single sourcing. These benefits can be considered as short term operational advantages, and as long term improvements. The long term benefits are:

- Supplier is responsible for the item
- Closer relationship with suppliers
- Joint improvement programmes to reduce cost
- Buyers' leverage is enhanced
- Focus on quality and efficiency
- Fewer vendor contracts
- More stable schedules.

Looking at single sourcing from the supplier's point of view, the advantages and disadvantages are:

Advantages
- Higher share of business
- Better opportunity to plan resources
- Confidence in continuing demand
- Opportunity for product and process development

Disadvantages
- More dependence on one customer
- Need to divulge financial information.

In the long term, the development of mutual confidence and a close working relationship will enable two partners in the supply chain to plan facilities to match the ongoing demand, focus on quality and delivery (just in time, JIT?), eliminate formalities (time and paperwork), and provide a service rather than simply a supply of items. The aim is to optimise added value throughout the supply chain for the benefit of both partners. This requires a commitment by both parties to collaboration over several years.

The short term benefits of single sourcing can be classified as:

- Volume per supplier is maximised
- Reduction in buying and paperwork
- Less vendor management
- Easier traceability for non-conformancies
- Single shipping route
- Consolidated deliveries
- Faster replenishment cycle.

To make single sourcing work in practice, management have to be convinced that the cost savings from these benefits far outweighs the potential costs of the risks identified.

What really is single sourcing? The obvious answer is 'only one supplier for each item'. This does not mean putting all your eggs in one basket, but it does mean relying on a supplier for a particular size of egg. This always gives the back-up that the size 3 egg supplier could be persuaded to provide size 2 if there is a big supply problem. Single sourcing also means empowering the supplier to provide the quantity, quality and delivery without fail. Strategies to support single sourcing are therefore:

- Ensure supplier profitability
- Ensure that demand is important to supplier (significant turnover)
- Partnership relationship
- Management and monitoring of performance.

Vendor appraisal

The traditional purchasing targets of good quality, price and delivery should be interpreted to mean total quality, long term total cost and consistent or JIT supply.

Quality is the responsibility of the supplier. Quality issues at the supplier interface are concerned with the provision of usable goods, delivery safely and on time, and accurate information including the customer's identity code, about what is sent, shipment by shipment, box by box.

It has been normal practice in some industries, to inspect goods arriving, on a 100% basis in some instances, and on a sample basis in more advanced companies. It is also customary for manufacturers to inspect goods being despatched. Items are therefore inspected on despatch from one company and arrival at the next either to ensure the integrity of the transport, or because there is a degree of mistrust. The integrity problem is relatively easy to solve, at least within one country. The mistrust is more difficult.

There are always stories of suppliers sending wrong quantities or wrong items, even though the consignment may have been inspected on despatch, particularly if the supplier is also the manufacturer. The integrity of the method of supply may also be questioned, more so if it involves international or long distance transportation. If a customer inspects a delivery and finds it defective in some way, a complaint will be lodged with the supplier and questions which then arise are:

* Who paid for a full inspection and when?
* Who separates the defective parts?
* Who pays for the return transport?
* Who owns the problem?
* Who pays for a vendor rating and quality control system?

In most cases the answers to the majority of these questions is the customer. If the supplier is not worried about this aspect of customer service (especially for a single source or accredited supplier), then getting the customer to carry out the quality checks seems to be a good cost saving idea. Only where the supplier's business is jeopardised by the quality situation is the matter resolved. The options for the customer are:

* Send the whole shipment back.
* Establish and circulate a league table for supplier rating.
* Charge the suppliers for the cost of inspecting their work.
* Communicate serious problems to their chief executive.

Table 8.3 Classification of suppliers

Class	Status	Inspection
Certified vendor list	First choice Close relationship	No inspection
Qualified vendor list	Select group of suppliers Business quality audited Offered expanding business	Minor audit inspection
Approved vendor list	Used on occasions Items not available elsewhere Prototypes, non-inventory items, office supplies	Guarantees or full inspection

The ability of a business to meet commitments to its customers depends on the quality of support from its suppliers. If that support is missing, the business must look elsewhere for better service.

Appraisal criteria

Audit is necessary as part of vendor assessment, to ensure that the excellence of supply is maintained. The purchaser's appraisal consists of three major areas:

- Technical – is the item fit for function?
- Supply – is the quality, quantity, timing, identification and cost acceptable?
- Support – is the supplier assisting in developments, queries, re-engineering?

The appraisal criteria in each of these areas can be listed and ranked according to importance – some features will be essential and others merely desirable, and suppliers can be classified by their past achievements against the set criteria. Results can then be used to rate the suppliers, as shown in Table 8.3 for the type of inspection required.

Vendor rating systems are operated by most major companies to ensure that their suppliers meet specified criteria. The final result is usually to categorise the suppliers into these classes:

- *Certified vendor list* – these are the small band of good suppliers who need little supervision and minimal quality checks. They should include the major suppliers.
- *Qualified vendor list* – medium supplier performance, but causing some non-conformance costs. Acceptable but needs work to improve.

- *Approved vendor list* – this includes all the other suppliers. Their performance history is either low quality, or unproven, so they require a quality assessment project to either upgrade their achievement or dismiss them.

Operating vendor rating

To establish and maintain vendor rating, a programme of vendor liaison has to be created, which should include not only a rating system but also frequency of inspection, inspection topics, focus and method, other visits and liaisons, the duration of the visits and the methods of recording and relaying the results. For those suppliers not given a top classification, the purchaser should report back on areas of concern, publicise to the suppliers the selection criteria and objectives and agree an improvement programme.

When a rating system has been specified, it has to be actioned. It is, therefore, useful to specify in advance the conditions under which suppliers would be abandoned or supported. It may not be possible in all cases to apply these conditions because of restrictions on supply, but it is useful to have prime criteria defined for the best working partnership.

The objectives of vendor selection are primarily to ensure adequate supplier performance and an agreement on financial rewards and penalties is very useful to make it work in practice. The supplier gets a premium if the rating is excellent and the purchaser gets a discounted price if the rating is poor. The overall effect of this can be to reduce the total quality cost between the two companies. As a spin-off from vendor selection further benefits accrue namely:

- Fewer suppliers
- Certified suppliers
- Objective performance data
- Long term business relationships
- Better understanding and dialogue.

Supply partnerships

Organising a partnership

The modern relationship of supplier and customer is considered in terms of long term collaboration, open relationships, mutual development, supply chain management and quality conformance. There are also some practical considerations including geographic location and flexibility. The process for developing a new supply partner is usually:

- Select vendor
- Assess expertise
- Accredit
- Agree supply type
- Test performance
- Set up agreement
- Monitor performance.

Monitoring is fed back to the supplier and maintained on a continuous basis.

The supplier base consists of a wide variety of different types of suppliers providing individual blends of service. These services should provide what the purchaser requires, but of course there is always room to improve in one way or another. Some suppliers are retained because they are essential (particularly where one part of an organisation is purchasing from another), others are retained because they are excellent. There is a continuing need to develop supplier relationships, to secure improvements and sometimes to seek alternative sources where suppliers are unable to meet the requirements.

When selecting possible vendors, both long term and short term considerations should be used. What information is there on the supplier, and do they, or are they capable of meeting the supply quality standards? It is important to ensure that the proposed demand level will not overload the suppliers, causing them to fail. If the style, objectives and business ethic of the supplier match the customer's, then a partnership is more likely to be successful.

The main reason for supply partnerships is to improve effectiveness for each of the participants through better inventory management. Improved mutual trust means better forecasting, less safety stock, availability of stock records to each party, shared risk and therefore low inventory. Successful partnerships agree on attainable targets which stretch both partners to give an equal share. The effects go outside inventory management into sharing development skills, operational and financial data and providing support for each other.

Pricing method

Because of the huge variation in market and trading conditions, the purchase price can be fixed in a number of ways:

- Market price – variable from day to day
- Fixed price – optional supply channel
                – mandatory supply channel

- Market adjusted price – using an agreed formula
- Fixed term contract – with discounts.

Each of these has advantages for supplier and customer in different markets and the choice should be one which both partners consider to be fair.

The ordering process

Planning orders

There are two separate types of procurement to be considered, namely purchasing and reordering. A large proportion of inventory control in a mature business will be the reordering of supplies from an agreed vendor and begins with communication that a further delivery is required. This situation is more about material control and scheduling than traditional purchasing.

If the purchasers can separate out those items where there are routine orders from those which are not likely to be repeated, then a different focus can be made in the initial negotiations.

The strategy for supply should take into consideration the overall commercial, logistic, inventory and added value services which are to be provided by the supply chain, and a variety of purchasing options can be considered, as appropriate. Supply can be obtained from:

- The open market
- A manufacturer
- A dedicated provider (specialist or distributor).

Each of these supply sources has different characteristics which should be examined to ensure their use is right for the items to which they are applied (see section on *Order types* below). Current practice is to set up a supply pipeline which goes through as few stages as possible. However, the nearer the original source (manufacturer) of the item, the less flexibility there is likely to be, the longer the lead times and the higher the supply volumes.

Order types

With the increased pressure for effective supply, simple one off purchasing processes are largely irrelevant. No longer does a buyer act as a dealer, seeking a number of suppliers, negotiating price and delivery and ordering a batch. This is far too time consuming and ineffective for the bulk of stock items.

Table 8.4 Analysis and application of supply orders

Order type	Advantages	Risks	Application
One-off orders	No commitment	No continuity of supply	Capital purchase
Bulk purchases	Traditional method for low prices	High inventory High risk	C class items
Annual contracts	Assured supply from awkward supplier	Need to forecast accurately one year ahead	Seasonal purchases (food)
Split delivery purchases	Lower inventory	Large stock cover or commitment	'Order' based supplier C or B class items
Rolling schedules	Assured supply low inventory, low commitment	Non-adherence to schedule, poor forecast adjustment	A class items, delivery more frequent than fortnightly
Just in time	'Zero' inventory pull system Short effective lead time	Delivery failure Unreliable forecast of demand	High volume limited range. Responsible suppliers

Ordering practices separate the purchasing aspects from the reordering aspects. Purchasing is strategic and carried out at intervals to set the structure, volume and price of supply. Reordering is carried out as a result of stock control and forecasting processes as necessary for managing replenishment. The two aspects require different skills and can be carried out by different individuals. Supply order can be broadly classified into:

- One-off orders
- Bulk purchases
- Annual contracts
- Split delivery purchases
- Rolling schedules.

An analysis of these types of orders and their application is shown in Table 8.4. The more organised companies use JIT or scheduled supplies. Unfortunately much of the software used in purchasing systems assume one-off orders will be placed, and so schedulers are forced to carry out extra work and make schedules look like split delivery orders. Rolling schedules and JIT supply are the only options which meet the real needs of both the supplier and the customer.

Order quantities

Types of order

If the aim is to minimise overall cost by keeping order quantities low, efficient stock monitoring and goods received systems are required. However, if this necessitates more frequent ordering, then the purchasing workload is increased. But an increase in inventory actually results from the delivery, not from the order, and the prime task must be to reduce the delivery quantity. Sometimes this is easy where reducing the order quantity is more difficult. There are several approaches, including:

- Supply scheduling (see next section).
- Multiple item deliveries.
- Source from suppliers who sell in small quantities.

A reduction in delivery quantities can be achieved in two ways, first by increasing the variety of items per delivery and second by increasing the delivery frequency. Given a varied delivery, the stock value is decreased in direct proportion to the number of lines delivered.

Consider the simplistic situation of a weekly delivery from one supplier who delivers one item in one week and another item in the alternate week. They deliver enough of each item for a fortnight each time, which provides an average of one week's stock of each. Instead it is agreed that half the quantity of each will be delivered every week. The delivery quantity is now enough for one week and average stock is then half a week's worth. The stock is halved by simply splitting the order quantity between the items supplied. As paperwork would be increased by the smaller deliveries, it is important to minimise the effect of the extra time spent by limiting this approach to 'A' class items.

Pareto based order quantities

The key decisions in stock control are when to order (discussed in chapter 7) and how much to order because the larger the order quantity, the higher the risk of excess stockholding. The order quantity depends on the value and usage rate of the item, and purchase order quantities should be related to the predicted requirements. The use of the Pareto technique described in chapter 3 provides the best balance between usage rate, costs and ordering effort.

The simple logic is to order A class item frequently to avoid stock value and order C class infrequently to avoid too many orders and deliveries. A typical order pattern would be:

- A class – order enough for one week
- B class – order enough for one month
- C class – order enough for 10 weeks

This of course is not dependent on the lead time taken by the supplier. Deliveries can be arranged every day even if the lead time is four weeks, it just means that there are about 20 supply orders outstanding all the time. The lead time does affect the safety stocks and therefore the overall stock level.

It is a gross assumption that one week's worth of stock is the appropriate order quantity for A class, but it appears to be generally correct across a wide range of businesses and markets. The main exceptions are fast moving goods where there is daily delivery and the equivalent balance in days is used as the order level. Companies where A items are delivered less frequently than fortnightly have large stores and unbalanced stocks, or they have several months' lead times for slow moving A items.

The effects of this ordering pattern on the number of line orders placed per year has already been discussed in chapter 3. If the above order pattern is applied to the example provided in Table 3.8, then the values shown in Table 8.5 result. The result, compared with monthly purchase for everything, is a reduction of stock from sufficient supply for 2 weeks' usage to enough for 1.325 weeks' usage (plus safety stock in each case) and a decrease in line orders of just under 5%. This gives more orders but less stock than shown in Table 3.8 and gives a better balance.

The ABC order quantities reflect the natural ordering practice for many inventory controllers, but the Pareto technique provides rules for controlling all the stock in an appropriate manner.

Table 8.5 Pareto order patterns for 1000 stock lines

Class	% moving lines	Order cover (weeks)	% turnover	Average weeks of value	Orders per year[a]
A	10	1	65	0.3	5 200
B	20	4	25	0.5	2 600
C	70	10	10	0.5	3 640
Total	100		100	1.325	11 440

[a] For 1000 moving stock lines.

Scheduling supply

The best arrangement for ensuring reliable supply and low stocks is the rolling schedule. It is commonly thought that parts which take a long time to arrive should be ordered in big batches. This is not necessary. Large savings in inventory cost can be made by receiving regular small amounts of the high turnover value items which have long lead times. For example an imported part may take 2 months from the order to receipt into stores. If there are 420 used per month, then an order could be placed every month for about 420. This would lead to a stock of 210 on average (plus safety stock). Alternatively, an order can be placed each week for delivery in 2 months time from the date of each order. This would mean a delivery quantity of 100 each time and average batch stock of 50. However it might be possible to place orders every day, in which case the batch stock would be reduced to an average of 10. This is a significant reduction in stock and cost without any effect on customer service. The number of orders per year may increase, but the workload can be reduced instead by using supply schedules rather than single orders.

Rolling schedules are not fixed on an annual basis because this may lead to a problem at the end of the year where there is either a shortfall or a large unwanted amount still to be delivered. The best schedule is an agreement between supplier and customer to the supply of an agreed quantity over a number of weeks. Outside this period the customer indicates the expected requirement so that the supplier can organise capacity and materials to cover it. Although the supplier would prefer a long fixed schedule, while the customer would like very short commitment, the scheduling technique has many advantages when operated properly with the full commitment of both supplier and customer to keep to the schedule. They include:

- Less risk for both customer and supplier.
- Reduced stocks for the customer.
- Reduced stocks for the supplier.
- The opportunity to plan supplies.
- Regular flow of materials.

A further advantage of scheduling arises from a regular flow of orders where, instead of placing an order or a new schedule, the old schedule can be used and extended. This enables a continuous supply of items at the average rate without the need for any purchasing action. There should be a simple agreement between supplier and customer that the schedule quantity rolls forward unless there is specific instruction otherwise, so that supplies are ensured. The only reason for changing schedules should be to adjust rather than create.

Not only does the rolling schedule embody the principle of 'no change – no action' (the classic management by exception principle) but it improves the stability of supply. Even if the supplier's lead time varies, the scheduled quantity is not affected, nor should the regularity of deliveries be affected either. A schedule is essentially a sequence of delivery times, not order placement times, so the equivalent of lead time for schedules is the inflexible front end of the schedule. The only change should be in the length of the fixed part of the schedule. In consequence as the lead time increases, the customer has a longer time ahead before the programme can be changed. Schedules solve many of the problems of ordering and mini-mise the risks.

To use schedules successfully, both supplier and customer have to be committed to the quantities agreed. There should also be no sudden changes in schedule outside the fixed part of the schedule. These are caused by late correction or overcorrection of demand or stock levels, and if not avoided, will give a stop–go ordering pattern. When usage changes the schedules should be modified gradually. Minor action early prevents a major problem later on.

Accurate systems are required to control scheduled orders. Running orders and continual deliveries are included on the schedule, and the cumu-lative delivery quantities have to be correct with a check to show which scheduled delivery is actually being received. A computerised technique is an advantage where there are many items to schedule.

The economic order quantity approximation

Where there are no other guidelines to determine the order quantity a the-oretical model can be used, the 'economic order quantity' (EOQ). This model considers batch size, Q, which results from balancing the cost of holding stock and the cost of ordering it. All costs are assumed either to vary with the batch quantity or to be invariant with the order cost. For an item with a cost of P and an annual sales of A, the cost of holding stock is given by $\frac{1}{2} \times r \times P \times Q$, where r is the financing factor. The half arises through averaging the usage rate. r is usually 20% to 30%, because it takes into account the cost of borrowing the money for the stock and the running cost of the warehouse. The order cost, S, contains the cost of running the order department, goods receiving and supply progressing and can be typi-cally US$100 per order.

Consider an item with a usage of 420 per month (5040 per year) and a cost of $35. Using typical values of $100 for the cost(s) of raising an order and 25% for the return expected on the stock investment, the ordering

Table 8.6 Balancing orders and stockholding costs

Orders per year (N)	Weeks between orders	Order quantity (OQ) (A/N)	Ordering cost (OC) (N × S)	Stock cost (SC) (0.5 × OQ × P)	Total cost (OC + SC)
1	52.0	5040	100	22050	22150
2	26.0	2520	200	11025	11225
4	13.0	1260	400	5513	5913
13	4.0	388	1300	1696	2996
26	2.0	194	2600	848	3448
35	1.5	144	3500	630	4130
52	1.0	97	5200	424	5624
104	0.5	48	10400	212	10612
Balance					
15	3.5	339	1485	1485	2970

options are shown in Table 8.6. The costs for various order patterns show a minimum for 13 orders per year, so this is said to be the EOQ.

At the bottom of the table is the actual minimum value, 15 orders per year. This has been calculated using the formula:

$$Q = \sqrt{\frac{2 \times A \times S}{r \times P}}$$

Substituting the values in this equation gives an EOQ, Q, of 339, which suggests that the item should be ordered 15 times per year.

Although the EOQ theory gives simple answers it generally gives too high a stock level and should only be used where there is no alternative. There are many reasons why EOQ gives the wrong answer in practice. These include:

- Order and stockholding costs are assumed to be fixed, but these costs should vary according to the stock and ordering situation from full to marginal cost.
- The demand is assumed to be regular and ignores batching and timing of issues which have a major impact on order costs in reality.
- Split deliveries and schedules do not fit well into the cost equations.
- EOQ ignores balancing of stocks which is the most important factor in manufacturing and many other cases.
- There is no allowance for coordination of orders for similar items where the order cost can be shared.

There are other minor problems which arise from the impractical assumptions made by the model. The application of this EOQ technique is therefore in the quick assessment of the stock and ordering situation. It is not appropriate for use in manufacturing, nor should it be used as part of a stock control system.

As can be seen in the example, the total cost is insensitive to the number of orders per year near the minimum cost. An increase of order frequency from 15 times per year to 20 times per year only increases the total cost by 4% but decreases the inventory by 26%. Even ordering every two weeks rather than every 3.5 weeks will increase the total cost by 14% but decrease the inventory by 43% as evaluated from the table.

Good techniques can reduce the order quantities below those given by the economic order calculation without increasing the costs. In many situations, batch sizes in the region of 60% of EOQ minimise the actual costs. The best answers are given by considering the practical situation and keeping the delivery quantities as small as can be sensibly achieved.

Other order size considerations

There are situations where the supplier imposes a minimum batch quantity, or a penalty for ordering small quantities. This suggests that there is a mismatch between the supplier's perception of a 'reasonable amount' and that of the customer's requirements, and is a sign that the supplier is not interested in the small amount of business which is available. Good liaison between purchaser and supplier could set up a rolling schedule to assure the supplier of continued custom, albeit for small quantities.

Minimum order quantities, as set by suppliers, are generally arbitrary round numbers. It is possible to erode them by a few percent in many instances and gradually move to the required quantity. Minimum order constraints on A class items should be studied energetically to avoid expensive stockholding. Where the minima are imposed on C class, then the solution may be to accept the constraint as long as goods are likely to be used in under a year. The minimum order quantity problem can be resolved in a number of ways offering different degrees of flexibility as shown in Table 8.7.

Where the supplier is a sole source, then more effort has to be made to resolve the situation:

- Customers may be persuaded to accept an alternative.
- A second source may be found.
- The item can be sourced through a third party (avoiding significant cost increase).
- The item could be destocked and the minimum quantity supplied direct to the customer.

Table 8.7 Minimum order constraints

Constraint	Initial policy	Long term changes
Minimum order value	Place order for phased delivery	Negotiate on volume of annual business
Minimum shipment size	Coordinate orders for widest mix of purchases	Change method of shipment
Minimum order quantity per line	Order for phased delivery	Schedule, or arrange payment when stock is used
Minimum delivery quantity per line	Share item with others (locations or competitors)	Renegotiate or look for alternative supply

Negotiation and understanding of the situation can sometimes lead to a sensible accommodation on both sides.

Apart from minimum order constraints there are other situations requiring specific policies. In many situations, the forecast will be considered to be acceptably accurate for a number of weeks ahead, up to a limit, commonly called the 'planning horizon' illustrated here for each.

Situation – by an Order policy:

- *Single warehouse* – Forward cover to horizon fixed by Pareto class.
- *Standard product* – Cover to horizon against forecast.
- *Distributed stockholding* – Coordinate supply to all stores.
- *Manufacture to order* – Buy and make required quantity plus scrap allowance.
- *Parts for assemblies* – Make balanced sets.

In manufacturing, batch quantities should be the same for purchase and production where possible. The actual batch sizes are often determined by the practical limits of capacity on a process, the overall process throughput time and the physical need to make a batch a handlable quantity. By coordinating the manufacture of like products the batch size of each can be reduced because the set-up costs are reduced. Thus a slow moving item should be produced in small batches along with similar fast moving items.

Purchasing processes

The stages of purchasing for a reorder are usually as follows:

1 Usage causes requirement.
2 System identifies need to purchase.
3 Need is assessed and actioned if necessary.
4 Purchase is initiated.

5   Order is produced.
6   Order is authorised (if high value).
7   Order is sent to supplier.
8   Supplier initiates supply.

If the workload in purchasing is too high to allow proper vendor development and liaison, ways must be found to reduce the ordering workload by attending to the eight stages and rationalising them, as follows:

Usage causes requirement

This is the initial step in the reorder process and the whole supply process is dependent on the information being accurate. The first issue is to ensure that the stock and demand information is correct.

System identifies need to purchase

Much time can be wasted searching through printouts, screens or even stores, to see what is really required. Systems should identify by exception, reporting those items considered to be ready for reordering. The reporting of this should err on the safe side by selecting all items which need ordering plus extra items whose demand patterns may lead to a high usage, a decision on which can then be taken by the inventory controller.

It is always useful to have a communication from stores which reports what they consider to be at risk. This avoids orders getting overlooked and adds weight to the supplier expediting process.

Need is assessed and actioned if necessary

The information provided by the system should initiate orders for a minimum of 85% of the items where orders are recommended. If this is not the case, then the stock control system is inadequate and requires improvement. The development of good stock control systems through forecasting or material requirements planning (MRP) is not a complex computer problem. Solutions are available for most situations, including sporadic demand and seasonal sales.

Purchase is initiated

The actual purchasing process can be a simple confirmation at the press of a computer key, or a time consuming discussion with one or more suppliers. Some companies even require a quotation for routine supplies! For fast moving items, the ordering process should be done in a flash once the order quantity has been decided. For most items the order quantities should be determined using the formulae discussed previously.

Order is produced

Order production is part of the normal computerised purchasing system and is printed automatically. Manual orders are to be avoided since they complicate the system.

There is a good argument for a petty cash purchase system where over-the-counter purchases of low value are carried out by an authorised person outside the normal purchasing system. This will avoid a large bulk of formal purchases and immediate access to the items required (often maintenance parts). Control over these purchases should be through the monitoring of monthly purchase values.

Order is authorised (if high value)

For capital goods and non-routine purchases, there has to be executive involvement in the commitment of company money. For routine supplies, this process only serves to delay the purchase and is a great disadvantage. The signing of reams of purchase orders by a senior person wastes their time and usually the orders are not investigated during this process. There should therefore be generous spend limits for buyers. Executives need to control the purchases by:

- Monitoring the value of order placed and outstanding order commitment weekly or monthly.
- Signing the major value orders (say 20 per week).
- Auditing the general order level by random checks each week.
- Setting targets for reduction of value of purchases relative to demand.

Order is sent to supplier

By post an order can take two days to reach the supplier, longer for overseas suppliers. The use of fax or more modern methods enable the supplier to receive the information immediately. The use of e-mail, wide area networks (WANs) or the data superhighway is important for ensuring fast ordering. This saves stockholding as well as improving communication. The well organised purchaser will not have to use the telephone for normal routine orders. The telephone is time consuming and should be reserved for supplier liaison and negotiation.

Supplier initiates supply

The information sent to the supplier should then be processed into their supply system. Any delivery problems should be relayed to the customer without them having to prompt. In situations where the customer is apt to alter the demand, there is a tendency not to report difficulties.

9

Forecasting demand

Options for assessing demand

Buying safety stock or purchasing a batch quantity means making a prediction that the items will eventually be sold. This is a crude type of forecast. In order to improve the stockholding (increase availability and reduce stock) the quality of the forecast has to be improved. If forecasts were excellent and the supplier could predict when the customers would want items, they could be obtained and delivered just when they were needed. Good forecasting means low stock. Poor forecasting means high stock.

Forecasting should be based on data which is accurate and appropriate for the purpose. This is often a problem and a forecast frequently has to rely more on balanced estimates of future sales rather than history. The data used to calculate the forecast include established demand patterns but ignore irregular demands for which stock is not normally held, such as scheduled customer orders for which specific purchases are to cover the demands, one-off promotions, sales campaigns or one-off very large orders (generally orders where the demand is more than four mean absolute deviations (MADs)).

The selection of the forecasting technique is a difficult decision and either the inventory controller selects the most appropriate one or employs a focus forecasting technique. In general, it pays to use the most sophisticated technique available because it is a better model for the demand pattern and gives better results. Forecasting techniques can be divided in many ways. One simple classification is between forecasts which use prediction and those which use history, another is between qualitative and quantitative figures, and another between intrinsic forecasting and extrinsic forecasting.

Considering these concepts in turn, historical forecasts rely on looking back to predict the future. Historical forecasting forms the basis of most inventory forecasting because it has sufficient detail, and demand for products usually changes continuously. However, it is obvious that looking back to go forward could lead to disaster and this is where predictive forecasting plays a role. People will predict an event based on understanding of future changes, rather than extending what has already happened. The problem with predictive forecasting is that it is often qualitative. A sales campaign will increase demand, but by how much? The qualitative prediction is marginally useful, but for stock control a quantitative assessment is necessary.

The difference between intrinsic and extrinsic is that the intrinsic forecasting is based on information which is available within the system. Extrinsic forecasting uses information which comes from external sources such as trade statistics or market information. It is convenient to split extrinsic information into leading indicators, concurrent indicators and lagging indicators, according to whether the information collected shows future, current or past activity. For instance, government statistics on housing starts is likely to be a useful leading indicator to a national roof tile manufacturer. The same information would be a concurrent indicator, showing what is happening to the general demand this month, to a company hiring digging equipment for house foundations, and for architects it would explain the level of activity which they had already experienced. Obviously, leading indicators are the most useful, even though the information provided is often qualitative.

The manipulation of forecasts is part of the inventory manager's art, since computerised systems are unable to provide the right interpretations when conditions change.

Types of inventory forecasts

Some types of forecasting are used only for specialised purposes. Of those used regularly by inventory managers, the following five approaches are the most useful and form the basis of stock control.

Market research

There is nothing like certain knowledge of what customers are going to buy. Firm orders are the ideal! Sometimes it is possible to obtain firm order commitment, but more often it is possible to gain only estimates from customers. Good liaison with the biggest customers or the biggest users of 'A' class items, can have a major impact on forecasting accuracy. If customers trust a supplier, they will provide accurate information, since they are in the best position to know their market and their likely order requirements.

An alternative would be to carry out surveys to discover trends in the market. These can be particularly useful in one-off investigations for new products or for determining the effect of different sales or promotional policies. It is a skilled and complex job to devise market surveys which give factual results for calculating the demand level. A market surveys is only cost effective as an ongoing exercise for a narrow item range where the volume is high.

Another method of assessing customer requirements is to ask the sales people. The risk with this is that sales staff sometimes confuse the difference between maximum potential sales and expected average sales. Consequently their estimates often have to be interpreted before use!

Market demand models

These are generally based on knowledge of the customer's markets. If the major factors which cause the demand can be identified, then a model can be created. The influence of each factor can be assessed and put into an equation, or the factors can be input into a simulation which can develop its own parameters.

Models are very effective where there are leading indicators and where there are simple deterministic relationships causing demand. Some models can be financial (expenditure on promotion versus increased turnover) others technical (item needs replacing every x years) or based on commercial factors.

Historical techniques

Forecasts can be based on historical data and then modified by the prevailing influences. The only exceptions to this rule are for new products and those which have a causal link with another forecast (via a model or MRP).

There is a wide variety of historical techniques to choose from. Some are simple and give adequate results, others are much more sophisticated and give excellent results in some circumstances. In general, the more sophisticated the technique, the more data is required to realise its potential, so established products can be forecast best using these. Also a weighted average technique will be better than a simple average. There are many approaches to historical forecasting, favoured by a variety of authorities. The basic models for these will be discussed in the following chapters.

Minimum stock levels

Many systems, both manual and computer have a 'minimum stock level' field displayed. The field is usually used for the 'reorder level' instead of for the minimum stock and is either a user input field or fixed by some useless rule (such as cover for one week).

With a large range of stock items, it is a very big task to maintain stock levels up-to-date. In practice, where the minimum is too low, the stock controller will increase the minimum level, but where the stock is too high, there is no sales problem and so fewer adjustments are made. As a result, the stockholding can easily increase gradually.

Minimum stock levels should be avoided and proper safety stock levels should be recalculated on a regular basis by a computer system. An area of development of stock control has been the improvement in forecasting techniques used by the professional stock controller. These should continue to be used to the maximum.

Demand patterns

The demand for stock items often follows a pattern which results from a variety of customer requirements for each item. There are two general approaches predicting demand patterns, first, treat the usage as a pattern and analyse the demand profile (an intrinsic approach) and second, investigate the causes of the demand and forecast the requirements separately (an extrinsic approach). The first alternative has the advantage that it is based on accessible information and is quick and easy to apply. This type of modelling does not normally predict rapid changes to the usage rate because it is essentially based on the historic demand pattern.

If a change in demand is understood, predictions can be made of likely usage and total likely requirements. This method should provide a more accurate answer, but requires thorough communication, knowledge and management.

In practice the first option, based on historical data, is used for the majority of items. High value items and critical items are usually treated by the second method, often through an informal system, although such systems should be avoided if possible in favour of extrinsic forecasting which will manage stock more efficiently.

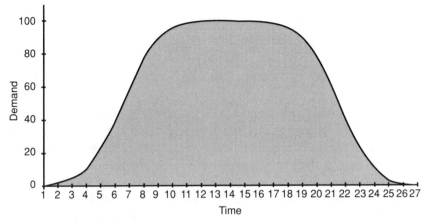

Figure 9.1 Product life cycle.

Product lifecycle

For the marketing and manufacturing departments of a company, a product has a lifecycle of three phases – design, development and manufacture. The end of manufacture is determined by marketing strategy and the introduction of a new more competitive product. This product lifecycle is shown in Fig. 9.1. It should be noted that product lifecycles are becoming much shorter with new technology. The product lifecycle for the mechanical typewriter was more than 25 years, for the modern word processor it is frequently less than one year.

For a service organisation, the need for stockholding may extend beyond the product manufacturing cycle since customers will retain products for many years after the end of manufacture and still expect spares to be available to support a service for their equipment. For some companies, products manufactured more than 20 years ago are still in use and being repaired as required. In some cases, service organisations have announced stop dates for service in conjunction with sales initiatives to sell customers new products. However this has frequently failed since third party maintenance organisations have seen an opportunity and taken over a segment of the business which is very lucrative.

The product lifecycle for the service organisation is different from the marketing and manufacturing lifecycle and can be divided into three phases:

- Introductory phase.
- Supply phase.
- Obsolescence phase.

Introductory phase

During this period, which should coincide with the design and development phases, the inventory planner will be working with the product project team when the customer service strategy for the product should be decided. To determine the stock requirement, the inventory planner will want to know the forecast demand and the possible geographical distribution of customers, which will enable the planner to determine the requirement and place orders for the items necessary to support the market strategy for the product ahead of the launch date.

Product launch is a critical period in the life of the item and it is essential to have sufficient to support the first customers. However if too much is bought initially, the profit can be seriously impacted because much of the initial purchase may never be used if too much 'just in case' inventory is bought. The planner must monitor carefully early life usage as the sales build up and adjust the ordering accordingly.

Supply phase

Established products have a steady or slowly varying demand pattern, and during this part of the product lifecycle major demand for stock items occurs. Inventory control is concerned at this time with maintaining availability. The ongoing requirement can be determined from historical usage data. Standard stock control systems with provisioning facilities can be used for reordering replenishment stock. At some point during this period, manufacturing may cease, which will create a 'last time buy' opportunity for distributors. Discretion is needed to ensure excessive 'just in case' inventory is not bought at this time.

Obsolescence phase

It is important that the inventory planner is involved as the requirement for the product declines. The task now is to prevent a build up of surplus stock and the stock control system should order fewer replenishment parts as the usage fails. However, a large proportion of stock will still be in place to meet the requirements of customer service. When the product demand ceases then all the stock should be withdrawn promptly. Companies find this a valuable exercise since it provides an opportunity to redeploy inventory in other countries where the product lifecycle is in its introductory or supply phase, and not its obsolescence phase.

During the obsolescence phase, consideration should be given to the relative importance of customer support and the high risk of having to write off unsold stock.

Causes of forecasting inaccuracy

People sometimes say 'you can't forecast this', a statement which is untrue. Everything is capable of being forecast, but some things can be forecast well and others are very difficult to forecast. To improve the quality of forecasts and therefore of stock control, it is useful to review what risks there are and what causes forecast inaccuracy. Typical pitfalls for the forecaster are as follows.

Inaccurate data

The whole of inventory management depends upon knowing how much stock is available so that the appropriate controls are maintained and action taken. Inaccurate data leads to poor availability, unreliable customer service and excess stocks. It is therefore imperative for stores to maintain records accurately and not just to maintain records by a weekly or monthly physical stock check. Records should be maintained accurately through a variety of techniques including automatic recording and checking, batch control, stores empowerment, perpetual inventory checking and, of course, bar coding. If records are not accurate, it is not worth improving inventory management processes until inaccuracies are small enough to be a minor influence on customer service.

Sales information rather than demand statistics

When considering inventory statistics, it is normal to discuss 'demand'. However the records of most companies focus on 'sales'. The difference is only important where the actual amount issued does not correspond to the amount ordered. This could happen in a number of ways, including sending boxed quantities or when there is a stockout. In the case of a stockout, the lack of sales could lead to the conclusion that no stock is required if no notice were taken of the actual demand.

Taking an extreme example: if an item has a demand over a 3 month period of 35, 45 and 40, but there is no stock of the item in the first two months and then a delivery during the next month, the sales records would give monthly sales of 0, 0 and 115. The actual fluctuations in demand are only about 10 whereas the sales statistics would give a much higher value to month 3. Professional stock records should therefore monitor 'demand' (order requirements at their delivery dates) rather than just sales.

Bias

Sales tend to use a forecast either to motivate the sales team by exaggerated optimism or to ensure that stock is available. As a consequence, projections from sales departments are notoriously high. Stock controllers have to resolve this basic issue either by interpreting sales data during forecasting, or by making sales (or marketing) departments more responsible for the accuracy of their forecasts. This can be done by apportioning financial responsibility for excesses.

Speed of response to change

It is difficult to get the right compromise between over-reaction to events and lack of response. One or two periods of high demand could signify an increase in market demand or simply a statistical variation. Either assumption will cause risk. Similarly when considering what time intervals to use for collecting data, if the time period is long (e.g. one month), there could have been major changes in customer demand profile within that time, or for short periods, demand can be erratic and make it difficult to interpret trends.

Poor assessment of supply capability

Fulfilment of increasing demand may be restricted by availability from suppliers. This can happen even with good suppliers who are used increasingly until they become overloaded and cease to supply successfully. This will result in forecasts which are not achieved. Supplier restrictions have to be understood by the forecaster. If a supplier cannot cope with peaks in demand, there is no point in including them in forecasts for peak rates.

Inclusion of extra demand in the forecast

Special offers, campaigns and fixed scheduled orders should not be included in the routine forecasts. They add variation to the demand rate which is not true random demand. Instead, there should be planned supply to meet the requirement which is assessed separately. For example, a demand pattern may be:

$$20 \quad 510 \quad 30 \quad 5 \quad 25 \quad 520 \quad 15 \quad 35 \quad 25 \quad 515$$

and the reason for this may be one major customer who requires 500 every 4 weeks plus a variety of customers who are taking about 15 per week. A sensible policy is to stock for the many small demands and to order in specially for the 500 to arrive just before it is required. The 500 would therefore not be included in the data for forecasting but as a separate type of demand in the item stock record.

The ability to forecast closely to the actual demand rate will depend on the selection of the method of forecasting. This can be done automatically by the computer from the historical models, but staff may also possess additional information which could be significant in forecasting future demand.

One-off abnormal demands can be excluded from forecasts by using a filter set at, say 4 MADs so that demands outside this range are treated as non-stock sales.

Shortage of data

For a 'good' historical forecast, sufficient data is required. It would be sensible to go back into history until the demand changed in pattern or level, but it is often difficult to assess this from the data. For seasonal sales, it is essential to have data for more than one year where, for example for fashion goods, the demand pattern may change within a few weeks.

The smaller the amount of information available, the less accurate will be the forecast. For example, the statement 'last 2 months' sales were 500' gives very little information on which to base stock levels. If the information available is 'the sales were 300 for month 1 and 200 for month 2' then this starts to give an indication of the demand pattern. If weekly demand data were available, it would be possible to see whether the demand was:

- only two customer orders or sporadic demand.
- regular orders each week or each day.
- decreasing demand.

This suggests that it is better to use more smaller time periods rather than large ones, unless the demands in the time periods become non-standard (for example, if demands on Fridays are different from the rest of the week, daily time periods would give cyclic demand with a peak each Friday). How much data is needed to make a decision? The answer is

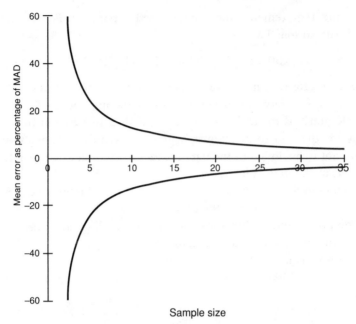

Figure 9.2 Effect of sample size on accuracy of mean.

indicated in Fig. 9.2. The size of the forecast error depends on how erratic the demand is, as indicated by the vertical scale. Figure 9.2 suggests that there is a reasonable degree of accuracy afforded by 12 periods of demand history.

Methods of improving forecasting

Feedback and monitoring

The simplest improvement to forecasting comes from ensuring that the forecasts are modified to reflect actual demand. This simply requires the re-evaluation of forecasts at the end of each period. New forecasts have then to be applied to inventory calculations to optimise the stockholding. Problem items can easily be identified in each period using a stock cover exception report. Feedback from the forecast error can show where the forecast technique is lacking and can identify when the forecasting method is not coping and needs changing.

Data aggregation

Statistically there is a major benefit from using as much data as possible by going back further in time or spreading wider geographically, locally and globally. The obvious risks of using old information are:

- Undetected population changes have altered the usage rates.
- Demand patterns have changed through modifications in user practice.
- Items have been substituted by alternatives or modified.
- Contracts or customer base have changed significantly.

If a wider geographic population is used, the risks are:

- Usage patterns differ regionally.
- Different demand patterns (or modification mix) exist locally.
- Data from different sources may not be compatible.

It is therefore a matter of assessment and balance as to the range of data which can be used for individual items. In general, the wider the information base, the more reliable the forecast. The accuracy improves as the square root of the amount of data included.

Comparative items

For a high usage item, random fluctuations are small compared to the forecast demand. Where demand is small, fluctuations can be large compared to the forecast. Alternatives to aggregation are first, to find a similar item which is expected to have the same demand profile and second, to vary the forecast for a minor item by scaling the forecast for a major item.

For example, consider two items in the same product range, one with a demand of 100 per month and the other with a demand of two per month. The forecast for the major item shows that the forecast has drifted from 100 per month to 110 per month. Assuming that the profiles are similar, the demand for the minor item should increase by 10% as well, to 2.2. It is much more difficult to detect confidently a change from 2 to 2.2 than from 100 to 110. It does not matter whether the two items change in exactly the same way. This technique works if the minor item is more likely to follow the major item than to remain static.

If there are a large number of items in a product range with low usage, comparison with the major items can be a good way of allowing for changes in the underlying level of demand. The forecast for each item can be based on individual historical demand and the resultant forecast multiplied by the relative change factor used on the major item.

10

Historical forecasting techniques

Basic forecasting techniques

Forecasting

Minimum stocks are possible when a good knowledge of demand exists. If demand were known accurately enough and far enough in advance, no stock would be necessary. On the other hand, if there is very little market information, high stocks are needed to ensure reasonable customer service. Forecasting is therefore one of the important aspects of stock control. It is a complex art which embraces many factors. There are two basic approaches to forecasting, first, assessing future market requirements and second, using demand history.

Assessment of demand requires knowledge of customers, products and background conditions that comes from discussion and evaluation with users farther down the supply chain through techniques such as market research, questionnaires and customer surveys. The strength of these techniques lies in providing general information on the levels of business and on the likely uptake of a new product or the phaseout of an old one. This information is invaluable where large changes in demand level are likely to occur.

These techniques need to be better managed to make them more acceptable as forecasting tools for each line item and inventory management needs to work with the sales departments to ensure that optimistic conclusions are not drawn, for example, higher demand, which then has to be second guessed for planning stocks. However, assessment of the market demand is an essential tool for the professional stock controller and the technique should be used for A class items and those at the start and end of the product lifecycles.

One method of determining customer demand is to assess potential sales. Many companies produce estimates, enquiry responses, or have contracts and formal agreements. Using a proven conversion factor on this information to calculate an expected average demand by item number can be the best way of estimating demand. Of course, a customer order could be considered as a market prediction, since there is always the possibility of the customer changing the date or quantity.

The second approach to forecasting, using demand history provides a convenient basis for predicting demand. For most items it is:

- Readily available.
- Detailed by item number.
- Generally reliable.
- Easy to use.

However using demand history is rather like rowing a boat. To go forward means sitting facing backwards and working hard. This is fine if there is no obstruction in the way but it is advisable to look round from time to time to see what is in front. In a similar manner, inventory controllers use history to determine the future. They have the problem that they are steering many hundreds of items at the same time and it is not practical to look round for each one to see if it is heading for collision. What is required is a general warning mechanism to detect impending disaster and a forward look for the more sensitive and important items.

Historical forecasting works for the vast majority of items and is therefore a basic tool of inventory control. The more sophisticated the forecast, the better the results, but there is always the need to keep abreast of real changes in the market so that forecasts can be overridden if necessary.

In practice, a variety of simplified models, based on major features of the demand pattern, can be used to meet stock control requirements for standard items. More sophisticated forecasting methods are applied to the non-standard items.

Historical forecasting methods are based on the manipulation of historical data in a mathematical way. This approach is fine for most items and can lead to excellent forecasts where the demand pattern is consistent. It is preferable to off-the-cuff estimates because it is:

- Consistent
- Rapidly calculated
- Frequently updated
- Based on facts, not reaction to pressures.

However it should be stressed again that the modification of forecasts in the light of additional information is also desirable.

Moving average

The most straightforward method of forecasting sales for the next period is to take the average sales for each preceding period, add them up and divide by the number of periods. The moving average technique is a formal way of calculating this for each period in turn, taking into account a fixed number of preceding periods (usually weeks or months). It is worth considering that financial accounts have a fixed annual cycle in which history builds up from the start of the year. This data is unsuitable for stock control, since a fixed number of periods of history is required. The moving average can be calculated from:

$$\text{average demand} = \frac{\text{demand for } N \text{ periods}}{N}$$

where N is the number of periods over which the average is taken. The selection of N is of prime importance in obtaining the best from the moving average technique. It is a compromise between the inaccuracy caused either by too little data (see Fig. 9.2) or by using old irrelevant data. Often the average is taken over 12 months (or 52 weeks) to avoid complications due to holidays or seasonality. However, it has to be remembered that by averaging 52 weeks of history, the data used is on average 26 weeks old. A lot may have changed since then! Therefore it could be safer to use a shorter moving average, but this may backfire since the data may not be sufficient to give a reliable average.

Take, for example, the demand history for three items over a 12 week period, as shown in Fig. 10.1. The average demand for all of them is 20.5 but the reliability of the number 20.5 as a forecast for the next week could be different in each case. In Fig. 10.1(a) the demand is in between 11 and 28, and 20.5 is quite a reliable forecast. For Fig. 10.1(b) demand is more sporadic and 20.5 is not such a good estimate – perhaps more data would be useful. In Fig. 10.1(c) there is a gradual increase in demand and 20.5 is rather low, so taking a shorter average would give a better answer in this case (i.e. a smaller number of periods). In general a 'suitable' number of periods is chosen for a whole range of stock items. This is not the best option but a practical solution. Calculations including many time periods are best where underlying demand is static, where there are large variations between demand in each period, where stock level is not too critical and the forecasting is allowed to run itself without much supervision. Reducing the number of periods in the average enables the forecast to cope faster with a changing demand level. This is useful where the demand is smooth or where

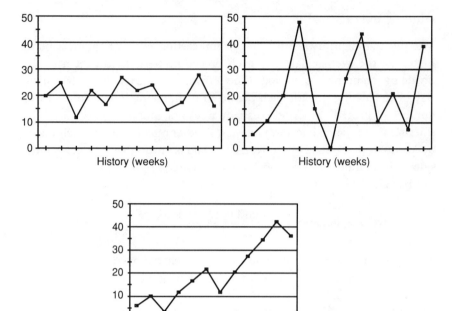

Figure 10.1 Historical demand patterns.

the sales in a period are dependent upon those in the previous period (autocorrelation).

Taking an average is really a method of segregating trends and random fluctuations. If the average includes data for a long period then any spurious effects are damped out, but this will also make the result insensitive to trends. Conversely a short moving average with small N gives good response to changes in market but is affected by fluctuation.

The histories plotted in Fig. 10.1 are provided with 6 week moving averages and 3 week moving averages in Fig. 10.2. Notice that the smoothest line is the 6 week average. The 3 week average is more variable, but it also responds better to the trend in Fig. 10.2(c). In practice a 3 week moving average contains rather a small amount of data, so if it provides the best forecast, then there is a major trend and a more advanced forecasting method would normally be used, or if not, then the 3 weeks could be split into days or half weeks to obtain a reasonable amount of data.

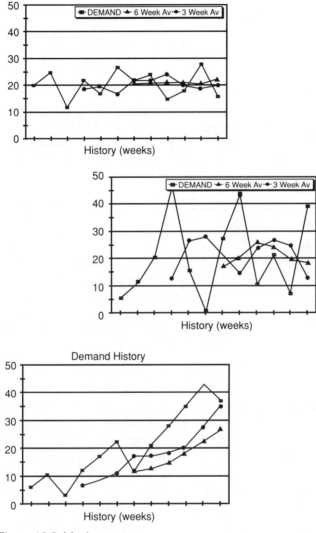

Figure 10.2 Moving averages.

Alternative calculation methods for a moving average

Calculating the moving average by summation month by month can be long winded, especially when dealing with 12 month averages. There is a shorter way of generating the new moving averages. For example, consider that the demand for an item over the previous 7 months was:

Month	1	2	3	4	5	6	7
Demand	17	11	15	12	14	15	?

The average demand, $A_o = 14$ for the first six months. In the following month the sales were 11. The new 6 month average demand A_N is therefore 13.

To calculate the old moving average, A_o the total 6 month demand of 84 was divided by 6. To calculate the new moving average, A_N the total 6 month demand of 78 was divided by 6. The demand for 17 in month 1 is not now included in the calculation, since it is older than six months. However, a simple way of calculating that $A_N = 78$ was $84 - 17 + 11 = 78$ so the new average can be calculated by working out:

$$\text{new average} = \frac{\text{total demand} + \text{new demand} - \text{oldest demand}}{N}$$

Even easier, and more practical, is to find the difference between the old demand data (to be discarded) and the new period data and divide the difference by the number of periods in the moving average. Then this can be added to the old average to give the new average, or:

$$\text{new average} = \text{old average} + \frac{\left(\text{new demand} - \text{oldest demand}\right)}{N}$$

Moving average is the most common forecasting method because it is simple to apply and easy to understand. However it is not the most favoured forecasting method because it has to be calculated from data extending back over the month by month period and equal weights are given to all periods. It is a matter of judgment whether the most recent information is more important in a particular instance. Often it is, and better techniques are required.

Moving average is most successful where the demand fluctuates widely because more sophisticated methods cannot identify demand trends and profiles in the presence of sporadic demand. Moving average is often used for C class items because the demand for these is often variable compared with the mean and it is more important to have a forecast which is generally reliable rather than one which requires some management.

Weighted averages

Benefits of weighted averages

The moving average technique suffers from two difficulties. First, it does not respond well to changes, as it gives equal importance to all periods and second, a large amount of information is required to recalculate the average each time.

An average which takes more notice of recent history and less notice of older data has major benefits. The use of a weighted average improves historical forecasting. There are many options for weighting the averages but there is one option which is better than the others and is used almost exclusively.

The technique of exponential smoothing avoids both the problems outlined above. It produces a weighted average which is based on all the information available (see Fig. 10.3). The calculation of a new average only requires the old average, the new demand and a weighting factor.

Exponential smoothing

The name exponential smoothing comes from the fact that contributions from history for an exponential curve go backwards in time. There is theoretically a contribution from many years of data, although this will become increasingly insignificant as it ages.

Forecasting with exponential weighting is the simplest of the professional forecasting tools. A new forecast demand or historical average, A, is obtained by mixing a portion of the old average with a portion of the new demand.

$$\text{new forecast} = \alpha \times \text{demand in last period}$$
$$+ \left(1 - \alpha\right) \times \text{forecast for last period}$$

The mixing portion, α, is the smoothing factor. A new forecast has to be calculated as soon as the data for the previous week or month (i.e. last period) has been collected. The new forecast then applies to the week or month starting at that time. However, in exponential smoothing (and moving average), the forecast for subsequent periods is the same as that for the next period. It is possible to project this forecast further ahead, but its accuracy will be poorer since it will not predict a trend in demand (up or down). The weighted forecast is easier to use when written down mathematically:

$$A = \alpha D + \left(1 - \alpha\right)B$$

where A is new forecast, B is the old forecast, D is the demand in the period just completed and α is the smoothing factor. A can be described as the 'average', which it is, although it is being used as a 'forecast'. The value of α is between 0 and 0.5 as discussed below.

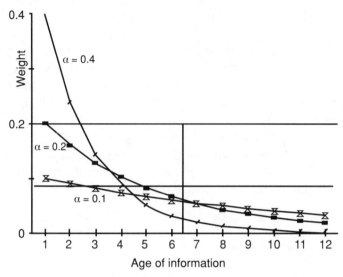

Figure 10.3 Contribution from each period to the average.

The smoothing effect of exponential smoothing is illustrated in Table 10.2. Like moving averages, this calculation simply analyses the demand to give an average plus the remainder which is treated as statistical fluctuation. The smoothing factor, α, defines the amount of notice which is taken of the demand in each period.

Selection of the smoothing constant, α

The nature of the forecast can be changed just as for moving averages by altering the value of the smoothing constant, α. Low values of α make the forecast consistent; high values make it reactive to change. Making it too high would render the result unreliable because it would depend too heavily upon demand in the last period. For effective exponential forecasting, the range of values which can be chosen for α is 0.1–0.4. If the demand varies slowly then a low value of α, say 0.1, is suitable, see Fig. 10.4.

If there are changes in average demand level to which the forecast should adapt, then a high value of α, say 0.4, is chosen. If the variability of the demand is high from month to month, then a low α is required to show up a stable average. Where the demand is smoother a larger value of α can be used. This responds more rapidly to trends in the average. If little is known about the item, or a value of α is required for a range of items, then

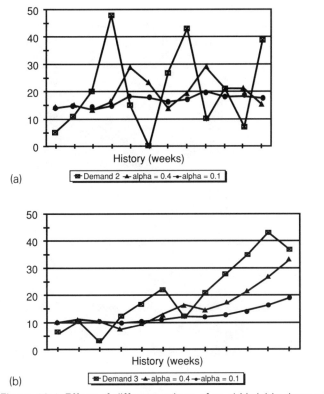

(a)

(b)

Figure 10.4 Effect of different values of α: a) Variable demand; b) Rising demand.

a median value of α = 0.2 can be used initially. The value of α is then reviewed at regular intervals and altered where necessary.

In Table 10.1 a sales pattern has been predicted each month using exponential smoothing with α = 0.1 and α = 0.3. The forecast for the next month has been worked out using the old average and the error. The success of the forecasting has been gauged by calculating the mean absolute deviation (MAD) in each case. The result indicates that α = 0.1 gives better answers in this particular case.

Alternative formula

This formula uses an alternative, more convenient method of calculation using the error in the forecast instead of the actual demand, which differs from the original form as follows:

Table 10.1 Exponential forecasting

Month	Sales	Forecast $\alpha = 0.1$	Error	Forecast $\alpha = 0.3$	Error
1	10	10.0	0.0	10.0	0.0
2	14	10.0	4.0	10.0	4.0
3	6	10.4	−0.414	11.2	−5.2
4	8	10.0	−2.0	9.8	−1.8
5	16	9.8	6.2	9.3	6.7
6	12	10.4	1.6	11.3	0.7
7	7	10.6	−3.6	11.5	−4.5
8	11	10.2	0.8	10.1	0.9
9	13	10.3	2.7	10.4	2.6
10	5	10.6	−5.6	11.3	−6.2
11	10	10.0	0.0	9.3	0.7
12	8	10.0	−2.0	9.5	−1.3
		MAD = 2.74		MAD = 2.88	

$$\text{new forecast} = \alpha \times \text{actual demand} + (1 - \alpha) \times \text{old forecast}$$

$$= \alpha \times \text{actual demand} - \alpha \times \text{old forecast} + \text{old forecast}$$

$$= \text{old forecast} + \alpha(\text{actual demand} - \text{old forecast})$$

so

$$\text{new forecast} = \text{old forecast} + \alpha(\text{error in old forecast})$$

This formula is used as frequently as the original but it is important to ensure that any negative signs are taken into account when using this version.

Contributions of history

The moving average takes data at full value until it becomes too old and falls off the end of the data table. With exponential smoothing, history is never discarded, it just becomes less significant. The forecast for this week, with $\alpha = 0.2$ is 20% of last week's demand and 80% of history. But that 80% was calculated from 20% of the demand of the week before and 80% of the previous history, i.e. last week is 20% of the forecast, the week before 16% (20% of the 80%) and previous forecasts are 64% (80% of 80%). Taking it back further, the contribution from 3 weeks ago is 12.8% and 4 weeks ago is 10.24%.

The contribution from each period to the forecast is shown in Fig. 10.3 going back 12 periods for $\alpha = 0.4$, $\alpha = 0.2$ and $\alpha = 0.1$. The forecast using

$\alpha = 0.4$ takes little heed of demands over about 6 periods old. The forecast using $\alpha = 0.1$ still has significant contributions from 12 month old data, and demands from as long ago as 18 months can affect the answer significantly.

Initialising the forecast

The fact that exponential smoothing depends on old history gives a problem of when to start the forecast. What should be taken as the 'old forecast' in the equation? There are two ways of solving this, namely estimate and history.

For a new product, or for the addition of a new item to the stock range, it is best to make an estimate of expected demand to start off the forecasting. This can be done by:

- Inspired estimate
- Model or replicate a similar item
- An arithmetic average of such data as is available
- Historical forecast

The effect of this estimate will be reduced at the rate shown in Fig. 10.3. It is not a good practice to start with zero as the old forecast since it takes too long to settle down, especially with low values of α.

Where there is some usage information, a large number of items to forecast, or inexperienced staff, then a historical forecast is better. If the item already has 18 months usage history, the technique is to work out a 6 month average for the demand up to one year before and to use this as the 'old forecast' to start 12 months of calculation using exponential formulae to bring the forecast up to the present day. Once the initial forecast has been established, the exponential average is self-perpetuating.

Improved values for mean absolute deviation

Any tabulated information can be weighted using exponential smoothing. It is important that the forecast is weighted toward the most recent data and that the variability data incorporates the most reliable up-to-date information to set the safety stocks. The calculation of MAD or standard deviation should therefore be exponentially weighted to allow its wide application in assessing forecasts and calculating safety stocks with exponentially weighted MAD and forecast error functions. Where standard deviations are used, the weighting has to be applied to the variance (standard deviation)^2. Updating of the MAD can be done most conveniently using the second version of the exponential smoothing formula:

$$\text{new MAD} = \text{old MAD} + \beta \left(\text{error in MAD} \right)$$

where

$$\text{error in MAD} = \text{actual absolute deviation} - \text{old MAD}$$

and β is the MAD smoothing constant and is exactly the same as α but using β shows that it is the MAD that is being smoothed rather than the forecast.

It is normal to take $\beta = 0.1$ when smoothing the MAD. The reason for this is that the MAD reflects the characteristics of the demand pattern and this would only be expected to alter gradually. If there is a sudden change in the variability of demand it is likely to be detected first through a change in demand level and forecast bias.

If the forecast for period 1 in the example is 36.2 and MAD is 8.5 from a previous assessment, the new MAD can be calculated either by taking the errors from the last six periods and averaging them to obtain a simple average or by taking an exponentially weighted average. The last two columns in Table 10.2 show the difference between the simple and exponentially weighted average.

Examples

1 – MAD forecast = 4, actual absolute deviation = 6
   – new MAD = 4 + 0.1 × (6 – 4) = 4.2
2 – MAD forecast (old MAD) = 4
   – Demand forecast = 17
   – Actual demand for period = 15
   – Actual absolute deviation = 2
   – new MAD = 4 + 0.1(2 – 4) = 3.8

This calculation would be the same if the actual demand was 19. Although MAD always has a positive value, the error in MAD can be positive or negative.

Exponentially weighed MADs, which give a better assessment of current variability, should be used wherever possible. An example using exponentially weighted averages and MADs is shown in Fig. 10.5 together with the exponentially weighted errors of the mean. The exponentially weighted standard deviation is insensitive to variances during each time period because a low value of β is used. This is a good feature for ensuring a stable

Table 10.2 Exponentially weighted MAD

Period (wk/mth)	Demand	Exponential forecast ($\alpha = 0.2$)	Error	Simple MAD	Smoothed MAD ($\beta = 0.1$)
1	28	36.20	−8.20	8.50	8.50
2	45	34.56	10.44	8.45	8.47
3	33	36.65	−3.65	8.77	8.67
4	39	35.92	3.08	7.96	8.17
5	19	36.53	−17.53	7.06	7.66
6	27	33.03	−6.03	8.57	8.64
7	22	31.82	−9.82	8.16	8.38
8	48	29.86	18.14	8.43	8.53
9	36	33.49	2.49	9.71	9.49
10	25	33.99	−8.99	9.52	8.79
11	18	32.19	−14.19	10.50	8.81
12	41	29.35	11.65	9.95	9.35
Forecast		31.68		10.88	9.58

calculation of safety stock, but it does make the task of setting the initial standard deviation (SD) more critical. The initial estimate used for MAD or SD will affect the calculation for many cycles and can give misleading results if not chosen carefully. An initial assumption of zero for MAD or SD is not good enough.

In order to compensate for a change in variance or to reduce the effect of initial conditions, the weighting factor can be increased for a few periods but β should be kept below 0.2.

Choosing the best forecast

Application of forecast types

The forecast models discussed previously give a variety of estimates for the future, some of which will turn out to be better than others. The problem is how to choose the best option. At one time, it was necessary to decide in advance what the characteristics of the demand pattern were and to choose the best model to use. Nowadays technology can choose the best forecast as long as the rules are defined.

The general principles which have been used to choose between the simple forecasting models are shown in Table 10.3. The forecasting models are described in this chapter and in chapter 11. The exponentially weighted models have the advantage that they are more sensitive to recent events and

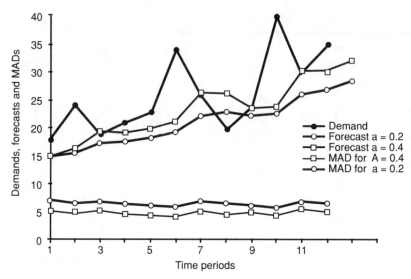

Figure 10.5 Exponential forecasts and MADs.

therefore are generally to be preferred over the simple averaging techniques such as moving average, regression or trend measurement.

Table 10.3 can be taken as an indication of which method is likely to give best results in any circumstances or it could be used to set the model for each item for forecasting. With the large number of lines in many stores, some gross assumptions have to be applied and rules can be suggested such as those shown in Table 10.4.

Focus forecasting

With the technology now available it is not necessary to make these assumptions. The approach is to calculate a range of options and then to select the best average. An assumption in making this selection is that the forecast which has been most successful recently will be the best one for the period being forecast. This assumption is open to debate but is the best working option. In inventory control the forecasts are usually for a short period ahead, making this concept more acceptable.

The question still remains as to which forecast can be considered the best. The best forecast is the most accurate. The most accurate forecast is the one which has the least errors (errors are the difference between the actual demand and the forecast!). Therefore, choice of the best forecast can

Table 10.3 Uses of simple forecasting models

Forecasting model	Demand characteristics
Moving average	Static demand level, irregular pattern
Regression analysis	Continuous trend, irregular pattern
Exponential smoothing	Variable demand, gradual changing level, and slow moving spares items
Base series	Seasonal demand, mature products with low variability
Double exponential	Varying trend in demand, low variability, product phase-in and obsolescence
Fourier	Product with long history, cyclic demand
More sophisticated techniques	Much history often needed, for longer term forecasting
Specific models	Based on market knowledge, potentially most accurate

Table 10.4 Simplistic selection of forecasting models

Pareto class	Model	Method of use
A	Double exponential	Monitor and use knowledge
B	Single exponential	Exception report and control
C	Moving average	Accept forecast
Other	Product specific	Supply to order

be made by finding the one which has demonstrably given the least errors recently.

The basis of a safety stock calculation is that stock is required to cover a divergence between the forecast and the actual demand (perfect forecasts mean no safety stock). Calculation of the safety stock is based on finding an average divergence (standard deviation or MAD) as one does in working out the accuracy of the forecast. The MAD is a measure of the forecast errors. If there are, say, three different forecasts for the same item, created in different ways, the way to choose the best is to calculate the MAD for each. The best forecast is the one with the smallest MAD.

Example

The demand pattern for an item over the past 12 periods is shown in the two left hand columns of Table 10.5. The forecasts in the three right hand

Table 10.5 Alternative forecasts for focus forecasting

Month	Demand	Forecasts			
1	28			Forecast	Error
2	45		Model	demand	(MAD)
3	33	Moving	12 period	26.8	5.9
4	39	averages	6 period	22.0	7.4
5	17	Exponential	$\alpha = 0.1$	28.3	9.5
6	27	smoothing	$\alpha = 0.2$	24.7	6.2
7	22		$\alpha = 0.3$	23.8	5.7
8	24		$\alpha = 0.4$	23.9	5.2
9	15	Double	$\alpha = 0.2$	21.9	5.4
10	18	exponential	$\alpha = 0.3$	21.8	5.0
11	28	smoothing	$\alpha = 0.4$	21.4	5.7
12	25				

columns were calculated as a result. By looking at the MAD column, it can be seen that the best forecast is double exponential smoothing with $a = 0.3$, and the worst forecast is exponential smoothing with $a = 0.1$. Therefore, the forecast to choose is 21.8. However at the next period end a different forecast might well prove to be even more accurate.

The use of simple MAD is reasonable but, as in the case of safety stocks, a better measure is readily available. This is because of forecast which was brilliant a few months ago, could now be rather poor, and it would be preferable to choose a forecast which has given reasonable results in the last couple of periods. For this reason a weighted MAD is used – in fact an exponentially weighted MAD (see section on *Improved values for mean absolute deviation*). This provides a more reliable method of assessing the potential of alternative forecasts.

Monitoring forecasts

Forecast tracking using cumulative sum of errors

The forecasting methods discussed will cope with most demands, but they need adjustment from time to time when the method is inappropriate for the conditions. We therefore need a monitoring process to enable prompt action to be taken.

Tracking signals provide a monitor of how well the forecast is performing. When the errors fall outside acceptable limits, the tracking signal identifies the problem or modifies the controls.

The Cusum technique provides a warning signal when there is a significant asymmetric shift of actual results from the forecast. It is simply obtained by cumulating the errors in forecast over successive periods. When the forecast is correct, the actual values are equally distributed about the forecast and cumulative error is small. Once the forecast diverts from the average demand, the errors build up period by period.

If the Cusum reaches an unacceptably high level, an error signal can be triggered and the forecast adjusted. The size of the error in the mean depends on the variability of the demand pattern and so the limit on acceptable forecast errors is set in terms of MAD – six times the MAD. However, for important items, values above four times the MAD should be investigated. When the Cusum has signalled a problem by a high error value, the reading should be reset to zero. Otherwise the controlled forecast will give a level which is consistently high.

Improving control – Trigg's tracking signal

When the Cusum indicates that there is a significant discrepancy, the forecast needs to be changed rapidly. For exponentially weighted averages, this means increasing the smoothing constant, α (see section on *Weighted averages*). Where there is a large variety of items this will result in many changes to be made. Trigg's method is a way of feeding back the Cusum information automatically and using it to modify α. Instead of generating the cumulative SUM of error, the tracking signal first works out the average error. This is done using the exponential smoothing formula on the forecast errors and is called the 'smoothed error', calculated from the actual errors, in the normal way

$$S_t = \delta e_t + (1 - \delta)S_{t-1}$$

where S is the standard error, e is the actual error and δ is the smoothing factor for the errors. The subscripts on the standard errors, t and $t - 1$, stand for the period just completed and the last period. Using the usual exponential smoothing of the MAD

$$\text{MAD}_t = \beta e_{t-1} + (1 - \beta)\text{MAD}_{t-1}$$

The tracking signal shows whether the forecast is good or not. It is considered good if it is not biased (i.e. not consistently high or consistently low). The tracking signal provides a figure of merit for the quality of forecast and is calculated by the ratio

$$\text{tracking signal} = \frac{S_t}{\text{MAD}_t}$$

where S and MAD are the exponentially smoothed values. The difference between the calculation of S and MAD is simply that S takes into account whether the forecast errors are up or down, and the MAD only uses the size (absolute value) of the forecast errors. Hence the biggest value that the ratio, the tracking signal, can have is 1, when all the errors are either positive or negative. The smallest value is zero where the smoothed errors compensate and are very small compared with the individual fluctuations in demand.

Trigg used this to give direct feedback into the α factor in the forecast. The argument is that an accurate forecast gives a low value for tracking signal and it needs little modification, so a low value of α is best. If the tracking signal is large, there is a significant error in the forecast and a high α factor is required to alter the forecast. The tracking signal can therefore be used as the smoothing constant for the forecast and this will enable an exponentially weighted forecast to be more self-compensating for each period. For this purpose, the tracking signal has to be a positive number and any negative tracking signals (TS) are changed to positive values. The forecast F, is then determined by

$$F_{t+1} = F_t + \left| TS_t \right| e_t$$

where e is the forecast error for the latest period and the value of TS is always positive. The tracking signal can be used for feedback into exponential smoothing or double exponential smoothing. It is also possible, but less beneficial, to calculate a tracking signal using normal averages. To make the tracking signal work properly the value of δ has to be the same as the value of β. Since both of these are usually set to 0.1 this does not present a problem.

There is a danger when using Trigg's method that the α factors can become very high (above 0.4) at which point it tends to overcompensate for small changes. In the discussion of exponential forecasting, it was stated that the value of α should be below 0.5. It is therefore advisable in practice to modify the formula so that values above 0.5 are not created.

The tracking signal is useful and allows a computer-calculated forecast to monitor itself and warn the inventory manager when it is producing bad forecasts. The tracking signal can be used simply as a warning flag. If the tracking signal is high, say above 0.4, then a manual adjustment to the forecast is required and an exception report can be triggered.

The tracking signal can be combined with focus forecasting, the optimum forecast being selected by choosing the one with the smallest smoothed MAD. If this forecast has a tracking signal below, say 0.4, then

Table 10.6 Forecast tracking

Period	Demand	Exponential forecast ($\alpha = 0.2$)	Error in forecast	Smoothed MAD $b = 0.1$	Smoothed error $\delta = 0.1$	Tracking signal
1	13	16.00	−3.00	4.00	0.00	0.00
2	1	15.40	−14.40	5.04	−1.44	−0.29
3	10	12.52	−2.52	4.79	−1.55	−0.32
4	17	12.02	4.98	4.81	−0.89	−0.19
5	11	13.01	−2.01	4.53	−1.01	−0.22
6	19	12.61	6.39	4.71	−0.27	−0.06
7	16	13.89	2.11	4.45	−0.03	−0.01
8	15	14.31	0.69	4.08	0.04	0.01
9	26	14.45	11.55	4.82	1.19	0.25
10	38	16.76	21.24	6.47	3.20	0.49
11	31	21.01	9.99	6.82	3.88	0.57
12	40	23.01	16.99	7.84	5.19	0.66
13	35	26.40	8.60	7.91	5.53	0.70
14	32	28.12	3.88	7.51	5.36	0.71
15	Forecast	28.90				

the forecast is acceptable. If the tracking signal is greater than the unreliability level (0.4), then the forecast with the next smallest MAD can be chosen. This is a simple, mechanical and effective way of obtaining good forecasts across the range of stock items.

Example

In Table 10.6 the characteristics of the demand pattern change after period 9. The forecast gradually drifts up and the MAD increases. The smoothed error which was small up to period 12 starts to increase. This is because the change in forecast lags behind the actual demand level. Consequently the tracking signal increases. It reaches 0.5 in period 14, which is quite quick. At that stage the forecast can be readjusted or an alternative model used.

Table 10.7, shows what happens when the tracking signal is fed back to change α as proposed by Trigg. The α factors are much smaller at the beginning. For the first few periods this is a result of the initial condition that the smoothed error is set to zero. The tracking signal stabilises and responds less to the low demand in period 7, which is sensible given the consistently larger demand in the previous periods. When the demand level changes, the tracking signal increases and enables the forecast to rise faster to cope with the problem. However this does not fully compensate and the

Table 10.7 Forecast using Trigg's tracking signal

Period	Demand	Trigg's exponential forecast	Error in forecast	Smoothed MAD $b = 0.1$	Smoothed error $\delta = 0.1$	Tracking signal
1	13	16.00	−3.00	4.00	0.00	0.00
2	1	16.00	−15.00	5.10	−1.50	−0.29
3	10	11.59	−1.59	4.75	−1.51	−0.32
4	17	11.08	5.92	4.87	−0.77	−0.16
5	11	12.02	−1.02	4.48	−0.79	−0.18
6	19	11.84	7.16	4.75	0.00	0.00
7	16	11.84	4.16	4.69	0.42	0.09
8	15	12.21	2.79	4.50	0.66	0.15
9	26	12.62	13.38	5.39	1.93	0.36
10	38	17.41	20.59	6.91	3.79	0.55
11	31	28.72	2.28	6.44	3.64	0.57
12	40	30.01	9.99	6.80	4.28	0.63
13	35	36.30	−1.30	6.25	3.72	0.60
14	32	35.52	−3.52	5.98	3.00	0.50
15	Forecast	33.76				

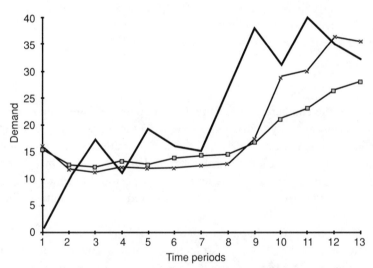

Figure 10.6 Effect of Trigg's forecast feedback. ——, demand;
─□─, exponential forecast, $\alpha = 0.2$; ─×─, Trigg's exponential forecast.

tracking signal still rises above 0.4. The forecast rose quickly to the new level as the α factor increased. It did this without the inventory controller having to intervene. The effect is shown graphically in Fig. 10.6, which contrasts the demand, the forecast with α = 0.2 and the forecast using Trigg's method. The data are taken from Tables 10.6 and 10.7 and represent a typical demand level change. The only problem in period 19 is that the tracking signal is still rather large. As the smoothed error continues to decrease because the forecast is again accurate, this tracking signal will reduce gradually. Although it is not the ideal answer since it lags behind the demand pattern significantly in a period of change, it is an improvement where there are many stock items and little time to manage them individually. However, it would be expedient to limit the maximum value of tracking signal fed back to 0.4 and to manually readjust when it exceeds this value.

11

Advanced forecasting methods

More forecasting tools

Improved forecasting techniques

Exponential smoothing is the basic model, which can, of course, be improved. It splits demand into two components, average, which is recalculated in each period and random fluctuations about the averages. The next stage of complexity is either to add a constant increase or decrease in demand, using double exponential smoothing, or a seasonal variation of demand, using base series forecasting. Alternatively, entirely different approaches to intrinsic forecasts can be used giving

- unweighted average trends (Regression analysis)
- profitability of changes in demand (Baysian forecasting)
- curve fitting (Fourier analysis)

or more complex methods.

The mathematics can become quite complex and are usually left to the computer. In general, the more complex the model the more history is required which is why simple forecasting has been acceptable until recently. Intrinsic forecasts are best for data collected under similar conditions, even when the demand is erratic. Then it is quite possible that exponential forecasting with $\alpha = 0.1$, or even long moving averages; are best. If a change from single to double exponential smoothing gives a significant improvement in the forecast for an item, it is possible that more sophisticated methods would work better, but there are fewer and fewer items benefiting from the better forecasting techniques. As a general rule forecasting can benefit more by improving customer communications to detect changes in the market than by going to highly sophisticated models.

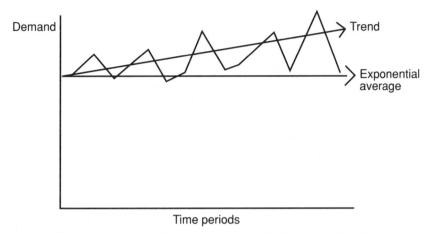

Figure 11.1 Improved forecasting through double exponential smoothing.

Double exponential smoothing

As stated in the last section, single exponential smoothing splits the demand into 'average + fluctuations'. This is improved in double exponential smoothing by analysis into 'average + trend + fluctuations'. In Fig. 11.1 the difference in forecast can be seen. The exponential forecast for the next period will show a weighted average, whereas by adding the average trend, the predicted rate of demand will be higher and the forecast will be better.

To understand double exponential smoothing it is best to go back to simple mathematics. Figure 11.2 shows a normal straight line graph given by the formula $y = bx + a$, with slope m and cutting the y axis at point c. The double exponential model is equivalent to this straight line, with y being the demand and x being time. Single exponential smoothing is like the line a above the x axis.

In double exponential smoothing the forecast of demand, F_t, is composed of two factors. $F_{t+1} = a_t + b_t$, with one factor arising from the movement of the average due to random fluctuations, similar to single exponential smoothing. The second factor calculates the trend of variations and adds a 'slope' factor. This has important advantages because the extrapolation to future periods continues, the increase or decrease being observed rather than assuming a static demand level.

In addition to this, double exponential smoothing can also compensate for lag in the forecast. It is obvious in graph (c) of Fig. 10.1 that the forecasts are lagging behind an increasing demand level (and will signal a

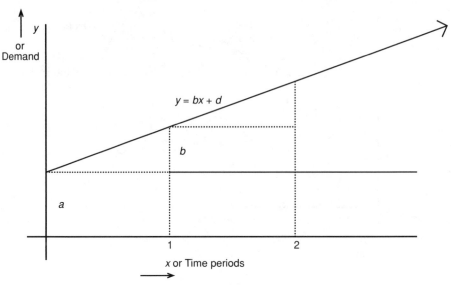

Figure 11.2 Double exponential model.

tracking error). Double exponential smoothing can bring the forecast in line with the demand once a linear trend has been established. This can be useful when there is a gradually increasing or decreasing average demand. The lag in an exponentially weighted average is simply given by the expression

$$\frac{(1-\alpha)}{\alpha} \times \text{trend}$$

where α is the smoothing factor.

From Fig. 11.2 it can be seen that the double exponential forecast for one time period ahead is $a + b$ (demand + trend). The forecast for two periods ahead will be $a + 2b$ (demand + twice trend).

There are several versions of the double exponential formula. The simpler version is Brown's model. A more comprehensive model is Holt's model, but this requires two smoothing constants (one for the base forecast 'a', and one for the trend 'b'). Holt's method does allow optimisation of the smoothing factors for selected items by iterative programming. The simpler formula is adequate normally and uses only one smoothing factor, which can be denoted as α since it behaves like the single exponential smoothing factor. The method of calculating values of a and b and the forecast is shown in Fig. 11.3. The general forecast for n periods ahead is

$$F_{t+n} = a + bn$$

1   *Find the Forecasting Errors*

Error = Forecast − Actual Demand

2   *Update the Demand Level*

Demand Level = Actual Demand + $(1 - \alpha)^2$ Error

3   *Update the Trend*

New Trend = Old Trend − α^2 Error

4   *Add the components*

New Forecast = Demand Level + New Trend

To set up the forecast requires:

Old Trend
Old Forecast
Actual Demand

Figure 11.3 Double exponential smoothing.

By this linear extrapolation of trend, a more accurate forecast of demand in future periods can be made if a linearly increasing or decreasing sales pattern is expected.

For items with widely varying demand patterns, the estimated trend can be misleading. A high or low value for the previous month's demand reflects too heavily on the calculated trend. If α is reduced to compensate, then the model becomes unresponsive. In this case it is better to take a longer sample period to even out the statistical variations and this is often done by calculating the average demand over the last three periods and using the averaged error to modify *a* and *b*.

The trend calculation is smoothed through the demands over preceding periods. Therefore, the greater the smoothing the slower is the trend to respond to change in trend. This causes lag – the amount of lag is shown in Fig. 11.1.

Table 11.1 Double exponential forecast, α = 0.2

Time period	Forecast F	Demand D	Error e	Demand level a	Trend b
				65	−5
1	60.0	55	5.0	58.2	−5.20
2	53.0	50	3.0	51.9	−5.32
3	46.6	45	1.6	46.0	−5.38
4	40.6	55	−14.4	45.8	−4.81
5	41.0	80	−39.0	55.0	−3.25
6	51.8	105	−53.2	70.9	−1.12
7	69.8	130	−60.2	91.5	1.29
8	92.8	155	−62.2	115.2	3.77
9	118.9	180	−61.1	140.9	6.22
10	147.1	205	−57.9	168.0	8.53
11	176.5	230	−53.5	195.8	10.67
12	206.4	255	−48.6	223.9	12.61
13	236.5				

Changing forecast methods to double exponential smoothing means that the initial values of a and b_{t-1} have to be estimated. This smoothing constant behaves like the smoothing constant for single exponential smoothing. The larger the value, the faster it responds to change, but the more unreliable is the value.

As the constant is squared for double exponential smoothing, selection of the correct value is more important. Lower values of α are chosen. In Fig. 11.4 the equivalent values are shown for changing from single exponential smoothing.

The method of estimating a and b depends upon how much information is available. If the demand for one period only is known or an estimate of it, there is little alternative to setting a equal to that value, b to zero, and using a high value of smoothing constant so that the forecast will respond rapidly to the actual trend. Thereafter, the smoothing constant can be reduced to a more stable value.

For an established product, the history is known and the values of a and b can be estimated accurately. In this case it is best to plot a graph and estimate the value of demand, F_t, and trend b_t. Then

$$a_t = F_t - b_t$$

These values of a and b can then be used as a starting condition for double exponential forecasting.

An example of double exponential smoothing is given in Table 11.1 and the effect of single period forecasting in Fig. 11.4 and 11.5.

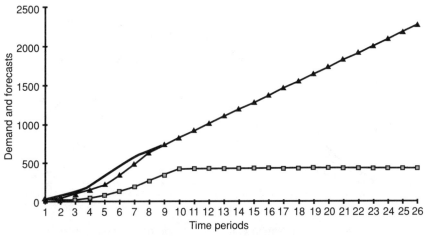

Figure 11.4 Forecasting several periods ahead. ——, demand;
—□—, exponential; —▲—, double exponential.

Forecasting for seasonal sales

For many products seasonal or cyclic demand patterns exist. This can be caused by natural seasonal factors, for example temperature, rainfall or the probability of thunder storms, or by other seasonal factors like holiday patterns, Christmas or other religious festivals, or financial year ends. These factors can be the overriding cause of demands, or a small reason for fine tuning the forecast. Seasonal demands can be masked by the sporadic nature of the demand for an item. Where this happens, as is often the case for C class items, it may well be expedient to ignore the seasonal variation and treat this as a random fluctuation too. Here a 12 month moving average is a simple solution, although the safety stock will, of course, inflate the service outside the faster moving season and cause more stockouts during the season.

Historical data

To use history to forecast seasonal demand requires data for at least one year. In fact the more years for which data are available the more reliable the forecast, unless there has been a change in the pattern of requirements of the item. This is usually a matter of interpretation, since the data is often distorted by specific, non-repeating factors to do with the market, price changes at irregular intervals or large contracts. Historical data often has to be interpreted to obtain the best forecast.

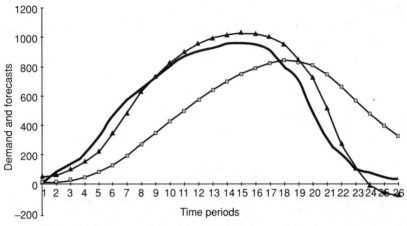

Figure 11.5 Forecasting with changing demand levels. ——, demand; —□—, exponential; —▲—, double exponential.

Generally seasonal data is collected in monthly buckets, although where Christmas dominates the sales, weekly information around November and December is often needed. The average demand for each corresponding month for three years or more is averaged so that an average monthly demand pattern emerges. It is useful to use weighted averages of demand to bias the data in favour of the most recent years. Where history is shorter, seasonal data can be used but there is less confidence in the results.

Base series technique

The classic technique for calculating seasonal demand is base series forecasting. This is a method of producing a seasonal demand forecast as an extension of exponential smoothing and assumes that the seasonal variation is known (from the data) and that forecast errors are due to changes in the average demand level or to random fluctuations. The process for updating the forecast is:

1 Take the new month's demand.
2 Deseasonalise it.
3 Compare with the forecast.
4 Update the forecast for the next months.
5 Reseasonalise the forecast.

Before the forecasting can start, seasonal demand factors have to be calculated and an initial forecast made for the average demand. The processes 1 to 5 are then carried out.

Example

Demand for July is forecast as 36. This is because the average monthly demand is 30 and history has shown that the demand next month is typically 20% above average. At the end of the month, the records showed that 33 were sold. What is the likely demand for August, which has typically proved to be 10% above average?

The demand in July should have been $30 \times 1.20 = 36$, but instead it was 33, which is 27.5×1.2. (The assumption is that the seasonal factor is correct.) The deseasonalised forecast should have been 30 but was 27.5. Therefore the forecast was too high and can be exponentially smoothed downwards (assuming α is 0.2 here) as follows:

$$\text{new deseasonalised forecast} = \alpha \times \text{deseasonalised demand} + (1 - \alpha)$$
$$\times \text{old deseasonalised forecast}$$
$$= 0.2 \times 27.5 + (1 - 0.2) \times 30$$
$$= 29.5$$

If the deseasonalised forecast is 29.5 and the demand for August is 10% above average, then the new expected sales for August are 29.5 + 10% of 29.5, or 32.45. The deseasonalised demand forecast of 29.5 can be used to forecast all future months by multiplying by their respective seasonal factors.

Other methods

From the wide variety of forecasting methods available, only the simplest have been described. Others such as regression analysis, Fourier analysis and Baysian forecasting are based on different assumptions about the demand pattern.

More comprehensive approaches such as Holt Winters and Wagner Within, which extend the methods to greater sophistication, can be used to give more accurate forecasts, but require more historical information.

As the quality of sales information is often inconsistent, the opportunity to use refined techniques is limited but the development of improved forecasting will make it easier for the stock controller to use better techniques in future. Meanwhile, application of exponential techniques is adequate for many purposes although changes will be required in other areas including the measurement of demand not sales, the segregation of stock from non-stock demand for an item and time bucket control. The application of refined forecasting will then provide useful benefits.

12

Material requirements planning – an alternative to forecasting

Avoiding uncertainty

When there is a clear picture of the demand for a particular item then stock can be minimised. No stockholding would be needed if demand were known precisely and far enough in advance because supply could be matched exactly with demand. For items outside the range stocked, the customer orders and receives the supply after a lead time during which the item is acquired. There may be some transient stock of these items because the supplier is asked to deliver in advance of the time that the customer requires it. The greater the confidence in supply, the shorter time these items can spend in stores.

The situation with non-stock supply is mirrored for raw materials in manufacturing and for production components in assembly. Stock is required to feed the process and firm plans exist for usage rate just as for scheduled customer orders and non-stock supplies. The approach to these types of inventory is to treat them as dependent demand rather than the independent demand discussed before. The contrasts between dependent and independent demand are shown in Table 12.1. The essential feature of dependent demand is that it is calculated from the demand of the next item up the supply chain; the aim is to have stock when it is to be used and no stock the rest of the time.

The stock profile for dependent demand is shown in Fig. 12.1. When the need is identified an order is placed. After the supply lead time, the delivery arrives. The items are then transferred or transformed for the customer and when this is complete (which takes the process lead time) then the item can be sent out. The stock is therefore only on site during the process lead time and the investment in stock is very low. There is potentially some safety stock included in case the customer order quantity turns out to be greater than that originally agreed. The safety stock here can be calculated from the forecast errors; in this case the 'forecast'

Table 12.1 Dependent and independent demand

Independent	Dependent
Stock level system	MRP system
Forecast	Calculated
Keeps stock	Buys as required
All lines separate	Lines coordinated
Reactive	Proactive
Good customer service	Very good customer service
High stock	Low stock

is the originally agreed delivery quantity. Ideally the safety stock should be zero.

A basic concept used in planning dependent demand is backward scheduling. This is simply the process of taking the delivery date required by the customer and working back to identify when the item needs to be ordered from the supplier. 'Offsetting by the lead time' is the jargon used for this method. In Fig. 12.1 the 'issues' should be made at the time requested by the customer. Moving back by the 'process lead time' gives the time to start on that order. If the system is working properly, this should happen just as the supplier delivered (marked 'receive' on the diagram). Going back further by the supply lead time identifies when to place the purchase order.

If this calculation results in a date which has already passed, the options are either to find a way of reducing the lead time, or to forward schedule from today's date. This forward schedule will then estimate a later customer delivery date so that the customer can be informed and the plan delayed.

Material requirements planning

The structure

There is a well developed technique for planning dependent demand called material requirements planning (MRP). This is a generally applicable technique for all types of dependent demand. The basic concept is to have stock when it is needed and to have none the rest of the time. With dependent demand, the size and timing of the requirements are known from the next level. MRP can therefore give good control. It is used extensively in manufacturing because there can be several levels of dependent demand in a product (assemblies made from subassemblies made from components

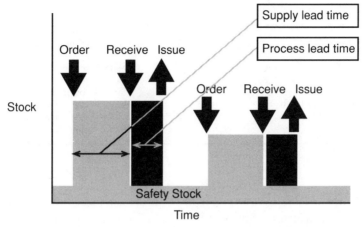

Figure 12.1 Inventory levels for dependent demand.

made from purchased materials). Here MRP is well developed and complex. In many inventory situations MRP is a much simpler process.

The MRP process

Material requirements planning is a useful planning tool outside manufacturing. Normal stock management treats each item as completely independent, unaffected by usage rates of any other item and the demand happens at random. The MRP approach is entirely different, based on:

- Interdependent usage rates between stock lines.
- Acquiring stock for when it will be used.

Many customers tend to use items together, so demand is not truly independent. For example, there must be a connection between the demand for hinges and doors, paint and paint brushes, nuts and bolts. In general business, there is no certainty that the sale of one item will be matched by the sale of the linked item. If a customer buys 100 bolts, there will be no automatic order for 100 nuts. They may already have a stock of nuts, or use some in threaded holes, or even buy from elsewhere. The situation will become clear after a while. For example it may work out in the long term that for every 100 bolts bought by a customer, 40 nuts of one type and 20 nuts of another type are required. This relationship could be consistent, because the customer's use is consistent, sometimes without identifying the reason. The demand for these items could therefore be interlinked and could vary as their level of business changes. It is therefore important for the items to be balanced in stock.

Table 12.2 Advantages of MRP over traditional stock levels

Stock level system	Material requirements planning
Treats each part individually	Deals with structure
Depends on demand history	Looks at future plans
Assumes average	Handles erratic demands
Aims to keep stock levels up	Holds stock only to cover demand
Priorities inflexible	Sensitive to priority changes
Mainly runs itself	Needs managing

MRP links in the inventory systems with the sales and financial plans of the company. It is a professional technique for a modern company and it requires adequate information, particularly:

- A master schedule of planned supply to customers for each product, projected far enough ahead to permit ordering of bought out items, raw materials and components. The planning horizon must be long enough to cover the procurement lead time for bought out items plus the total manufacturing time. Often this is more than a year.
- A well defined bill of materials showing purchased items and manufactured components.
- Lead times for the purchase or manufacture of all parts.
- Accurate records of stock, work in progress and on-order parts.

MRP enables parts to be scheduled for the day that they are required. The advantages over the traditional stock level systems are shown in Table 12.2.

The two main attributes of MRP, namely 'time phasing' and 'structure' are now being applied in a wide variety of stocking situations outside manufacturing. If it is known or can be forecast when major customers want their supplies, any business can create a master schedule, with the consequent reduction in stocks.

MRP is a business planning tool as well as a material supply calculation. The planning of MRP is shown in Fig. 12.2. The process, known as manufacturing resource planning (or MRP II), starts with a long term plan, which identifies what range of items are offered, how fast, with what market focus, and all the policy decisions to enable inventory control to operate successfully. As a result of this long term process, action should have been taken to ensure that the resources (people, plant, systems, logistics, organisation) are available to make the policies work successfully. This long term plan stretches ahead at least 6 months and, typically by two to three years,

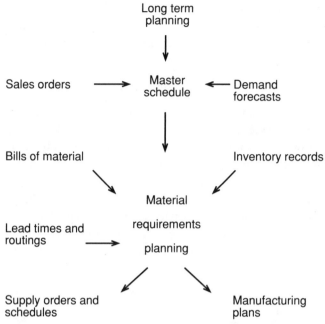

Figure 12.2 Material requirements planning.

depending how quickly the market is changing. Long term planning is usually carried out by product group rather than by product. A master schedule is generated from the long term plan and is simply a list of what the business is going to send out, allowing for the real constraints.

MRP logic

The MRP backward scheduling process starts by identifying the customer requirement in the master schedule. Once the output to customers has been decided, the focus is on achieving it. If there is already stock, how far along the schedule will it last? If there is no inventory, then what has to be purchased, the item itself or are there components which are needed? A bill of materials is required to cover this. At the same time, information on how long the supply will take and where it will be done – the lead times and routings – is needed. All this information is fed into the MRP calculation.

The actual MRP calculation is very simple. First it batches up all the requirements within each time bucket (week or day) so that there is no differentiation between different customer orders. Stock is deducted from this total requirement. If the stock is sufficient to cover the demand, the same calculation is carried out for the next period. If the stock is insufficient then delivery (order receipt) is required. The process is simply shown in Fig. 12.3.

initial stock + receipts − demand = closing stock

If closing stock is above zero (or safety stock) repeat calculation for next time bucket. If closing stock is below zero (or safety stock) then create more receipts by ordering more. This then causes extra supply for orders and demand for that component. This netting off process is carried out for all periods of output first, and then it is repeated for the supplying level (components or assemblies) if there is one. The result is a list (or a series of lists) of what has to be done by when; a list of when to place and receive purchase orders and, where appropriate, a manufacturing 'work to' pro-gramme for each production area (called 'manufacturing plans' in Fig. 12.2).

Bill of materials

Classic stock control deals with each item individually, independent of all others. However, for most inventory the usage of one item is linked to that of others, e.g. printer cartridges and paper. Inventory controllers need a way to acknowledge this link and to save the work of two forecasts where one will suffice (see Table 12.1). This can be done using an inventory bill of materials.

Traditionally, a bill of materials (BOM) is the list of parts, ingredients or materials needed to produce or assemble the required end product. The BOM for a product is not simply a parts list. It contains more information. Most products are packaged in some way and the box, pack-aging material, pallet and even the documentation is part of the full BOM. The BOM can also contain essential tooling for production. This is a useful way of planning for all the resources which are necessary for the process.

In manufacturing, an assembly is made up of components. If they are all bought out and then given to an operator to assemble and pack, then the BOM is as shown in Fig. 12.4. This is a flat BOM. If the assembly and packing are separate, then the BOM is as shown in Fig. 12.5. In a supply chain this would correspond to different suppliers of items which are to be sold together in the end. Time delays between processes need to be recorded and therefore add levels onto the BOM.

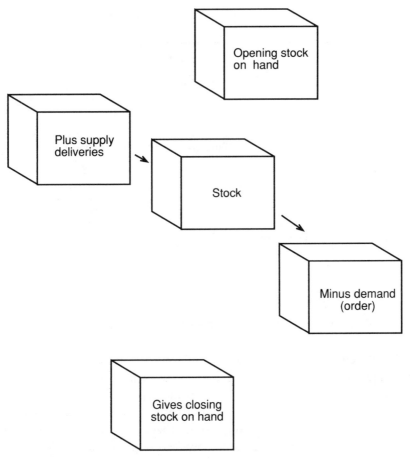

Figure 12.3 MRP logic.

Planning the structure of supply

In many situations there is a balance between the usage of different items. For example DIY stores know that someone fitting a door hinge will probably need screws. They will either have the screws added to the pack, or remind the purchaser that they are needed. Many activities result in the use of a variety of items and if this usage is a fixed ratio, or even if the ratio of usage is consistent on average, then the ratio can be expressed in a loose BOM. This could be used by a supplier providing a range of items to major customers. If the causes of use are understood, a planning (statistical) BOM can be created even though there is no manufacture involved. The use of this tool cuts down the amount of forecasting, since one forecast will provide the information for all the items included in the BOM.

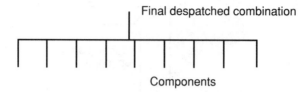

Figure 12.4 Balanced demand – bill of materials.

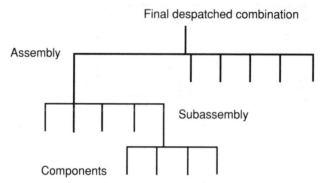

Figure 12.5 Structured demand – bill of materials.

The structure usually consists of two levels, the same as a just in time (JIT) manufacturing bill. The top level is the demand profile for the collection of stock items. The level below is the dependent demand for each of the item codes. Since the ratio is not entirely fixed, the bill has to be monitored and corrected according to the changes in the ratios caused by usage. This situation is similar to a manufacturing procedure where there is major variation in process yield or there is product scrap.

The demand for each item within the statistical BOM will vary for individual customer orders. This means that actual safety stock of each item will consist of two components, safety stock against the error in the overall forecast and individual line safety stock against the variances in the demand BOM. In practice, the statistical BOM is an excellent and underutilised method for distributors to give enhanced service where customers buy a range of items.

Master planning

Master scheduling

Improvements over classic stock control can also be made by considering exactly when the demand will occur. For example, if there is a known demand for 50 this week, 100 next week and 60 the week after, a know-ledgeable traditional controller could work out the average demand as 70 per week with an MAD of 20 and then decide on the appropriate safety stock. A different approach would be to identify the demand in each week as 50, 100 and 60 and arrange supply accordingly. To do this requires a new type of record. Instead of the normal record of what has been issued each week, the requirement is for a list of what will be issued each week, showing individual weeks (or days). This is a master schedule. The result of using a master schedule instead of a review level is that the inventory is kept lower and large fluctuations in demand can be accommodated. There is still the option to keep some safety stock, calculated from the inaccuracy of the schedule each week.

The master scheduling process for receiving orders converts the existing demand forecast into firm customer demand by placing the order on the master schedule. This is a management function not a clerical function. The master scheduler is responsible for allocating the quantity forecast for the customers. Normally, particularly outside the demand lead times, the forecast quantity will be greater than the total quantity ordered for a par-ticular time period (usually a week). However, when orders are received which exceed the forecast, decisions are required.

If the forecast is 100 per week and the first order is for 70 in the first week, then there is no problem. If a second order is received for 30 for the first week it can also be supplied. However if a third order is received for 50, for that week, this will have to be met from the forecast in the second week.

The logic of allocation is that the forecast is filled up by real customer orders to the forecast level each week. However, if the customers require more, then the master scheduler has to decide whether to:

- Provide later delivery or part shipment to the customer.
- Reallocate goods already promised to another customer.
- Overload the forecast and cause panic buying and inefficiency.

The last option is commonly used but should be avoided since it makes supplier relations bad. There is sometimes the option of increasing capac-ity by overtime, but the material supply chain may not be able to react to short term schedule changes.

The master plan in its simplest form is a list of how many of each item the company is intending to supply to the customers each week (or each day for JIT business), see Table 12.3.

Master planning takes the forecast and plans to supply parts to meet this demand at the right time. The scheduled receipts are the delivery date of

Table 12.3 Master schedule

Item	Week 1	Week 2	Week 3	Week 4	Week 5	Week 6
Type 1	1000	500	2300	600	1300	1300
Type 2	300	150	450	300	350	600
Type 3	3000	2000	3000	3000	3000	3000
Type 4	1500	~	~	~	~	~
Type 5	~	~	~	~	~	~
Type 6	~	~	~	~	~	~

current firm outstanding supply orders and the stock on hand is the inventory which results if no action is taken. The first three columns in Table 12.4 show the current situation. The remaining three show what has to be done to meet the forecast demand.

The planned order receipts are calculated to minimise stock and at the same time give the information necessary to supply forecast. The resultant stock level is shown in the column headed 'Projected available', i.e. the stock levels resulting from these actions. With a 4 week supply lead time, the order release (last column) has to precede the order receipt by 4 weeks. If a planned order release occurs in the current week, then an order has to be placed. Alternatively, the planned order release column is used as the time phased gross demand for the items on the next level down on the BOM, or as a supply schedule for purchasing.

Placing the order results in the order quantity being deducted from the planned order release and the 'planned order receipt' being transferred to a scheduled receipt. If a safety stock is appropriate instead of letting stock go down to zero, then the stock on hand can trigger a planned order receipt at the safety stock level. A safety stock of 1000 would mean that stock would be too low by the end of week 5. The planned order receipt and release would therefore be required a week earlier.

Delivery promising

When the customer orders are received, the forecast quantities should be available. The calculation of projected available and planned order release enables the supply to be planned. This calculation accepts the forecast and works out what supply is required to meet it. The amount of the item remaining to fulfil outstanding demand can be calculated in a similar way by looking at the actual orders instead of the forecast. The allocation of the projected available stock to real orders enables customer delivery times to be estimated using the MRP logic. The process is shown in Fig. 12.6.

Table 12.4 Planning supply – planned order release

Week	Forecast	Scheduled receipts (a)	Stock on hand	Planned order receipt	Projected available	Planned order release (b)
CURRENT			500		500	0
1	3000	5000	2500		2500	0
2	4000	5000	3500		3500	5000
3	2000		1500		1500	0
4	2000	5000	4500		4500	5000
5	4000		500		500	5000
6	3000		−2500	5000	2500	0
7	1000		−3500		1500	0
8	3000		−6500	5000	3500	0
9	4000		−10500	5000	4500	0
10	2000		−12500		2500	

[a] Assume a fixed batch size of 5000.
[b] Supply lead time is 4 weeks.

Creating an order book according to Table 12.4 gives the availability shown in Table 12.5. This shows the actual customer demand at the present moment but orders are constantly being received, so the situation may well change within a few minutes. The scheduled receipts and the planned order receipt are the values calculated from the projected available stock shown in Fig. 12.6. The stock on hand and the available to promise columns are calculated from the real customer demand column.

In week 1 there are 5000 received and 2600 issued to customers. The stock which was originally 500 is therefore 2900 by the end of that week and these could be used to fulfil customer demand. However in week 3 the available to promise is down to 100 and the supply lead time is 4 weeks. If more than 100 out of the 2900 are issued in week 1, then there will be a shortage in week 3. The delivery promise for quantities up to 100 can therefore be in the current week; the delivery promise for up to 1900 can be in week 4 and greater quantities can be promised either up to a maximum of the forecast or in excess of the cumulative forecast, as long as the forecast is changed to match the demand. This change has to take account of the fact that the large demand is probably part of the forecast and therefore the forecast for other weeks will need to be reduced.

The two sectors, projected available and available to promise, ensure that customer demand is managed in line with supply. It avoids difficulties which result from accepting orders on a fixed or short lead time which then cause emergency sourcing or failing to provide delivery on time for other customers.

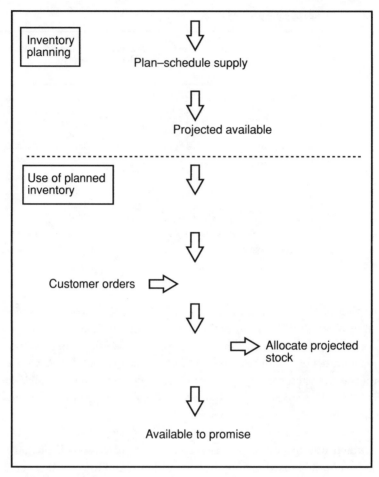

Figure 12.6 Inventory planning forecast.

Batch sizes

The master schedule is an excellent tool for managing and monitoring stocks, especially where there is variable demand which can be estimated. It enables lower stock to be held than can be achieved by statistical stock-centred methods. It also links in well with supplier scheduling. Getting the best out of master planning does require more effort because of the regular reviews and reforecasting. However in a pure stock environment, it is only necessary to use MRP on the A class items and items where the demand is understandable and variable. The triggering of MRP by this schedule enables companies to forecast for whole ranges of stock with a single

Table 12.5 Available to promise

Delivery promising	Customer orders	Scheduled receipts	Stock on hand	Planned order receipt	Available to promise	Planned order release
Current stock			500			
Week 1	2600	5000	2900		2 900	
Week 2	6600	5000	1300		1 300	
Week 3	1200		100		100	5000
Week 4	3200	5000	1900		1 900	5000
Week 5	1200		700		700	
Week 6	600		100	5000	5 100	
Week 7	4000		−3900		1 100	
Week 8	100		−4000	5000	6 000	
Week 9			−4000	5000	11 000	
Week 10			−4000		11 000	

evaluation and give a balance of stock for customers who require coordinated supply.

In classic MRP, a number of batch rules have been developed. As many of these are based on 'economic batch sizes', which produce batch quantities which are unbalanced and too large for customer order requirements, then the same logic can be used for MRP order quantities as for normal stock control. The three main alternatives are:

- Lot for lot
- Fixed batch size
- Period cover.

Lot for lot is the first choice because it leads to a minimum of inventory. The rule is to bring in the balance required in each time period. In Table 12.4, the scheduled receipts for weeks 1, 2, and 3 would have been 2500, (3000 forecast demand less the 500 stock) 4000, 2000 and the projected on hand would always be zero. This gives a variable supply quantity which is inconvenient is some circumstances.

Fixed batch sizes are used widely to illustrate MRP. They are overused in practice since changes in demand level should be reflected in changes in order quantities. Fixed batch sizes are convenient to use but are only essential where there is a restriction in vessel size (distribution or manufacturing). Batch sizes based on price breaks are the result of too simplistic an approach to costing.

Period cover is a way of taking the demand for the next few periods and batching it up. It is the same as lot for lot but with several time periods added together. Again the order quantity is variable and the stock level is significantly higher. This technique reduces the number of orders and increases their size at the expense of higher inventory. The application is therefore to C class items where this is the most beneficial option.

13

The future – inventory and logistics

Review of techniques

The science of inventory management has gradually developed, not because the theories have leapt forward but because pressure has become greater. Three major factors have been the increasing focus on customer service, the more competitive environment and the wider use of information technology. The development of computers and communication methods has enabled most companies to move from crude item level control to good control and aggregate inventory management. The computer does not forget items, lose records (at least not when used properly) or calculate wrongly, and it can immediately highlight situations for action. It gives up to date management totals and analyses which can be used by inventory professionals. The achievement of tighter targets has largely been the result of better communication and simple controls.

Basically, the key tools have been the exception reports of 'these stocks are too low' and 'these stocks are too high', which keep the stock broadly correct as long as orders are placed at approximately the right time. The use of ABC analysis, dynamic review levels, forecasting techniques and just in time (JIT) supply enables inventories to be improved significantly. Effective implementation of these techniques has been slow because conditions did not encourage it. In the past, supply lead time was unreliable so that review level calculations could not be relied upon. Now that suppliers are expected to guarantee reliable delivery the calculations are easy.

Techniques described in previous chapters have proved invaluable in meeting the demands of the new market environment. Once a company has established these, further improvements can be considered. There are other techniques, built into some advanced software, which are waiting to be used more widely and more advanced inventory control techniques will surely show significant advantages for the user. It has been known for decades that Newton's law of gravity is not quite right and Einstein's theory of relativity is nearer the truth, but Newton's law is good enough for those of us who are not space travellers or who move below the speed of light. Similarly in inventory control, the review level calculations are not exactly correct and delivery frequency does affect the safety stock. Better formulae are readily available, but from observation, stocks are so far out from the ideal that a small amendment to the safety stock level is not going to alter the real inventory. Similarly Gaussian statistics, simple availability calculations and independent demands are all concepts which are questionable theoretically but which work in practice.

More advanced forecasting methods provide better answers when the future can be relied upon to mirror history (see chapter 11). Crude tools like ABC analysis could be replaced by tools which take into account the potential different stock policies for high value slow movers and low value fast movers. Or ABC could be developed into a differential value analysis, which gives an individual class for each item code and optimises every inventory line. This evaluation will take place when companies have exhausted the current methods and realise that it can provide benefit.

Logistics

Improvement in provision or collection of information has given better operational and cost control. This has particularly led to a great change in the way that delivery costs are viewed. Transport costs were originally taken as a fixed percentage overhead on operating costs and the responsibility for these costs was of no concern to the inventory manager, except in the most general way. However, as supplier and customer delivery costs are now seen largely to result from inventory policies, they become a stock control issue. For example, a low inventory may result in many deliveries which increase the transport costs. This is good for inventory value, but may not be optimal because of the transport costs. A change in transport arrangements may allow operating benefits to be maximised. It is only by considering the two costs together that this can be achieved and so transport costs become a variable part of the inventory costing. This additional aspect has been added onto the complex inventory balance and forms the basis of logistics.

There was a time when logistics was concerned with the military ordinance of goods. Logistics took on the air of freight distribution and then encompassed warehousing. Now logistics can be considered as the domain of the distribution and warehousing functions linking the manufacturing processes and the end user. Logistics has become a science.

Logistics management and materials management present new challenges for the inventory manager, who has to maintain and balance distributed stocks, additional inventories, manufacturers' work in progress, supply partnerships and other new dimensions including distribution costs. The need for this integration results from the changes in market structure or distribution environment given here:

- Stockholding has become strategic warehousing.
- Transport has become more reliable and faster.
- Operations have become international.
- Transport and distribution costs are being optimised.
- Subcontracting is the norm (especially for warehousing and transport).
- Pressure on inventory cost is greater.

The development of supply chain strategy means that stock controllers from suppliers and customers should be sharing information and inventory in order to optimise the operation of the supply chain. The supply of items is characterised by universal underlying themes. The most successful supply chains are integrated, consistent, fast moving (short lead time), service oriented and customer specific.

Distributed inventory

The management of a supply chain requires stock to be held in different locations, even in different operating companies. The inventory controller can then optimise the stock across many locations. The structure of the supply chain can be altered to meet the needs of the market and the overall balance between availability, stock investment and operating cost can be improved by selecting the number of stages (echelons) in the supply chain and the number of stock sites in each echelon. This can be established most simply by modelling the balance. Generally it is more economical to hold stock as near to the start of the supply chain as can be allowed by the demand lead time.

Once the distribution structure has been decided, the proportion of stock held at each inventory level has to be calculated. A policy of holding all stocks in a single central warehouse at source has the advantage that the investment in stock is minimised. This advantage results from improved forecasting, higher demand rate and reduced safety stocks. In addition there is better physical stock control because of the larger stores operation.

If stock is all held by the end user or local stores, items become available for use more rapidly. However local stores normally means many more stock locations (stock keeping units, SKUs). This means a large duplication of stockholding, poorer control, higher stock investment and greater risk of obsolescence.

The compensating factor in deciding where to hold stock is the distribution cost. In general, it is considered that:

- Fast delivery costs more.
- The greater the distance, the greater the cost.
- The smaller the load, the higher, relatively, is the distribution cost.

However, there are other considerations to make, including:

- Regular shipments cost less than one-offs.
- Distribution costs are largely loading and unloading, mainly independent of distance.
- Consolidation saves cost and is easier for small, regular deliveries.
- Small loads need smaller transport and facilities which are cheaper.

The complexity arising from these latter considerations makes transport costing difficult. When considering just in time supply (see chapter 4) one of the principles discussed was continuous improvement (Kaizen). This approach has to be used when reducing distribution costs and, therefore, inventory. However, to have a significant effect on the cost balance, bold leaps have to be made in distribution structure. The size and frequency of shipment has to be altered dramatically. This usually means a change in means of transport to provide smaller units faster. The savings in inventory and warehousing costs should be greater than any marginal increase in distribution costs.

The correct balance of stockholding and the location of the stock is different for individual items, depending on their usage rates and values. Low value, high usage items are required nearer the point of use. High value items with occasional usage can be held centrally.

The whole inventory system can be considered as a pipeline for supplying the fluid 'stock' to the point of sale (or use). The input to the pipeline is a continuous (scheduled) flow. For low usage items the flow ends at a reservoir, the national stores. From there the flow is dispensed as required. For medium usage parts, the pipeline can split and end in the regional stores where there are smaller storage tanks to fulfil sales demand. For the high usage items the pipeline can continue to the point of use. This analogy highlights some interesting aspects of the supply chain, and poses the questions:

- Should the system required be a stock control system or should it be a supply flow system?
- Which parts of the supply chain are best served by scheduled supply and which by batch ordering?

- If stocks are held at one echelon, do they need to be held at others as well?

For fast moving goods, the central warehouse can act as a reservoir replenishing the local storage tanks. It is better if the central stores is used purely as a trans-shipment point for these items. The pipeline is then connected to the distributed stocks. The central function becomes stock management coordinating ordering and supply and ensuring consistent delivery.

Slow moving items can all be held at the central stores, as long as there is the means of withdrawing them and transferring them to the final customer within an acceptable time. This reduces the overall stockholding of these items. The resultant stock is structured with each item SKU having a specified stock level based on its usage and value.

Multilocation supply chains are used commonly in some industries. Using an integrated stock information system, duplication and proliferation of stocks can be avoided. Distribution warehouses in different localities should be treated as a single level where all stock is considered as holdings of a single stores apart from the safety stocks. The aim is to keep the appropriate amount at each location and if supply from the originator or manufacturer is rapid, then a central stock location can be avoided, with a corresponding reduction in stock. Careful supervision of delivery time is necessary to avoid duplication of stockholding at several locations.

The degree to which a company can take advantage of these techniques is limited by their organisation and systems. Often it is necessary to make improvements as a two-stage process:

- Introduce the 'stock to stock' system to obtain the control.
- When control is good, reduce the stock at each stage in the system by improving the inventory control techniques.

Stock distribution

Techniques available

For control of inventory in the distribution chain, the costs have to be minimised and service maximised. Distribution systems are neither 'pull' nor 'push' methods of stock control. Such methods imply that either distribution (pull) or manufacturing (push), are entirely responsible for the organisation of distributed inventory. Modern techniques centre around an accessible database of information, the creation of a master schedule and the replenishment of distribution sites based on total needs, requirement lead times and supply availability. Three techniques available for controlling inventory distribution are:

- Transportation algorithm (for minimising distribution costs)
- Fair shares (for allocating available stock)
- Distribution requirements planning (DRP) (for planning supply).

The transport algorithm optimises the cost of transportation. Both the fair shares and DRP techniques are concerned with the quantity of stock to be shipped through the distribution network. These techniques are used when there is good information fed back from all the distributed stores and central decision making for the allocation of stock.

Assessment of demand is more accurate for the system as a whole, since the larger volume has a smaller mean absolute deviation (MAD) (the MAD only increases as the square root of the volume). Although this means that the forecasting should be done by a central planner, the effects of local situations have to be included. It helps to have integrated systems.

Transport algorithm

For an inventory planner controlling stock in several regions, the issues to resolve are:

- Where the stock is to be held.
- What level of stock to hold in each region.
- How to maintain supply at the lowest cost.

The first question is answered by working out the general distribution and inventory cost balance from the factors just discussed under *Distributed inventory*. The second issue is easily resolved using the classic inventory management approach described in chapter 7. The third issue is to ensure that the transfers are carried out most economically within the framework arranged by the first and second issues. This is an optimisation process called the transport algorithm. It can be based on a range of assumptions and one illustration is presented here.

The cost of providing items at a remote warehouse depends largely on two factors the item cost and the distribution cost. There is a unique cost for each item at each site, and SKUs provide the information required to evaluate distribution needs. The stock in some distributed warehouses is likely to be larger than necessary, particularly if the warehouses are the delivery points for outside supply, whilst others need to be supplied with stock to meet demand. Stock therefore has to be redistributed at the least cost from one location to another at the same echelon of the distribution network. If the costs of distribution between each of the stores is known and the current inventories and required inventories are known, the cheapest way of achieving the redistribution can be calculated using the transport algorithm. This is particularly important where transport is expensive, or where there are significant differences in transport costs, as found in a multinational supply operation.

Using a simple approach usually results in a fixed distribution structure with replenishments being carried out along set routes without regard to the specific supply needs for that distribution. Often a better solution is to organise the distribution of stocks to meet the particular conditions. The transport algorithm considers the problem as a whole. It optimises the cost of trans-shipments by determining which stores with surplus stock should support those stores with shortages. Excesses and shortages can be the result of using fair shares analysis.

The transport algorithm is a stylised application of linear programming. The normal assumption about transport cost is that it is unique between each pair of stores and that the cost increases directly as the number of units moved. The distribution cost for 10 items is 10 times the cost of moving one. This is a good assumption for heavy or bulky items.

The method of working out the transport algorithm is to set up a redistribution as an array and (by convention, starting from the top left) to allocate the available surpluses to the deficits. The distribution costs of this reallocation can then be calculated. The technique is to change one distribution route which will cause others to be changed, to retain the balance of distribution. If the marginal effect of this change has reduced the overall distribution cost, it should be adopted for as many items as possible. If not, an alternative routing can be tried. By starting at one corner of the distribution array and trying the changes in turn, the least cost solution can be found.

In evaluating the optimum solution, the real costs of distribution should be considered. This cost can be reduced greatly, for instance, by synchronising delivery of different items and by the use of transport for return shipments.

Fair shares

Fair shares analysis approaches the regional stock issue as an inventory control problem, assuming that distribution costs are not the prime concern. The purpose of fair shares analysis is to distribute stock through the supply chain to alternative distributed storage. Stock is not held at a central warehouse but distributed to regional stores nearer to the customer demand. An implicit assumption is that the supply to the network can be accomplished by short lead time orders. If the lead time is longer, the planning aspects of DRP have to be used.

When stock is delivered into a distributed stores network, it has to be reallocated to the various stores in the most equitable manner. The sensible approach is to cover priorities first and then share out the rest to meet future demand. Fair shares analysis is a method of achieving this using a standard procedure. The logical process is to identify the safety stocks in the supplying warehouse and in the other warehouses and then to allocate the total available stock to meet the requirements in proportion to the demand rate.

The fair shares allocation of stock normally distributes stock in the following sequence:

• Customer back orders unsatisfied.
• Expected demand within supply lead time.
• Safety stock at distributed stores.
• Safety stock if required at supplying warehouse.
• Allocation of remaining stock in proportion to demand rate.
• Redistribution where a stores stock exceeds its share.

If stock is insufficient to meet the full requirement at any stage, it is allocated proportionately. Stock can be spread correctly through the supply chain using this logic. The actual stock transfers used to achieve this distribution can be optimised through the transport algorithm. A simple example of the use of the fair shares logic is shown in Tables 13.1, 13.2 and 13.3.

A manufacturer sells items in batches of 400 to a central supply warehouse. This warehouse holds some safety stock, but distributes the items out to three regional stores. Estimates for the usage at the stores are given in Table 13.1 together with the safety stocks. If a delivery of 400 arrives

Table 13.1 Stock and forecast demand

	Safety stock	Current stock	Week 1	Week 2	Week 3	Week 4	Week 5	Week 6
North stores	20	70	50	40	70	20	50	50
South stores	46	200	100	140	60	110	150	150
West stores	12	10	20	30	0	50	20	20
Total	78	280	170	210	130	180	220	220
Supply warehouse	37	200						

Table 13.2 Fair shares priority allocation of stock

Location	Average demand	Lead time (weeks)	Demand in lead time
North stores	46	1.2	56
South stores	118	1	119
West stores	23	0.4	10
Total	187	4	185

	Stock	Balance
Current stock	280	
Supply warehouse stock	200	
Current delivery	400	
Total	880	880
Shortages	0	
Demand in lead time	185	695
Stores safety stock	78	617
Warehouse safety stock	37	580

Table 13.3 Fair shares apportionment of stock[a]

Stores	3.1 Weeks demand	Demand in lead time	Safety stock	Current stock	Delivery required	Total stock
North stores	143	56	20	70	149	219
South stores	366	119	46	200	331	531
West stores	71	10	12	10	83	93
Total	580	185	78	280	563	843
Supply warehouse			37	600	−563	

[a] Number of weeks cover = 580/187 = 3.10 weeks.

from the manufacturer just at the start of week 1, the total stock available is 400 plus 200 central stock plus 280 at the regional stores (=880). Table 13.2 shows the lead times and develops the supply chain calculation. There are no back orders, so the first allocation of stock is to cover regional demand in the lead time (185), then regional stores safety stocks (78), then supply warehouse safety stocks (37). Deducting these from the total available leaves 580 items to be split up across the regional stores. This can be done either by netting off the individual forecasts until the stock is exhausted (where these are accurate and contain real orders). Alternatively the average week's cover can be calculated. The total demand is 187 per week, so 580 will last 3.10 weeks on average. Multiplying the demand by 3.10 gives the fair share for each store. This is shown in Table 13.3.

Each store is then given a fair share consisting of:

any customer back orders + demand in the lead time
+ safety stock + 3.10 weeks' demand − current stock

This is shown in Table 13.3. All the stock and delivery to the supply warehouse has been allocated, apart from the safety stock (37). If one regional store already had sufficient stock, then either the excess can be transferred, or that regional store (and its stock) can be excluded from the analysis and a fair share worked out amongst the others.

Distribution requirements planning

DRP provides planners with a complete forward schedule for all despatches, arranged by date order by warehouse. Each stores at the demand end of the supply chain generates a forecast for demand which is used to identify the orders they are likely to place on their supply warehouse (supply). The forecast is for demand well in advance of the normal replenishment lead times. Since the supply is planned to meet the requirement well ahead, the demand rate may have changed so the resultant replenishment delivery may be too early or late by the time demand is met.

The delivery requirements calculated by the netting logic of DRP show the latest shipping dates associated with each warehouse requirement (see Table 13.4). DRP is an extension of the MRP process for distributed warehouses. The netting logic used for DRP is exactly the same as for MRP (see section in chapter 12 on *MRP logic*):

Table 13.4 Example of distribution requirements planning

		Current	Week 1	Week 2	Week 3	Week 4	Week 5	Week 6
North stores	Forecast		50	40	70	20	50	50
	Stock	70	20	80	10	90	40	90
	Supply			100		100		100
South stores	Forecast		100	140	60	110	150	150
	Stock	200	100	60	0	90	40	90
	Supply			100		200	100	200
West stores	Forecast		20	30	0	50	20	20
	Stock	10	90	60	60	10	90	70
	Supply		100				100	
Supply warehouse								
Demand upon	North stores		100	0	100	0	100	?
supply	South stores		100	0	200	100	200	?
warehouse	West stores		100	0	0	0	100	0
	Total		300	0	300	100	400	0
	Stock balance	200	300	300	0	300	300	300
	Deliveries		400			400	400	

- The future requirement is split into time periods of a week or a day.
- The expected demand each time period is identified.
- Starting with the current period, the demand is netted off against the available stock.
- If the remaining stock is above the safety stock then this quantity is carried forward to the next time period and the calculation is repeated.
- If the available stock is less than the safety stock, then a receipt is required. The quantity requested depends on the batch rules being used.
- The request generates a supply order a lead time earlier. (The lead time corresponds to the picking, packing, transfer and receiving time.)

By using DRP, supply planners can receive a complete forward schedule for all despatches, arranged by date and by warehouse. Transport must be made to fit the schedule. DRP builds on that master scheduling and looks at demand far past the normal lead time. The purpose of DRP is to produce a time-phased shipping requirement for the suppliers.

DRP is a method of planning replenishment for the distribution network. The logic closely follows that of MRP. Like MRP the supply chain is planned by backward scheduling from the customer. This requires the establishment of fixed supply lead times and batch quantities. If this is organised by the manufacturer, the batch size can be large and the intervals in distribution correspondingly long. This of course adds stock throughout the supply chain making the use of a just in time approach with very short lead times a great advantage.

An example is shown in Table 13.4. In this table the items are distributed from a supply warehouse to three regional stores. The stores have given their estimates of demand (shown as 'forecast') and from these the stock and demand pattern needed to keep the 'projected on hand' positive are given in the 'stock' and 'supply' rows. The item is distributed in batches of 100 in this case. This is the same calculation as for 'projected on hand' and 'planned receipts' for MRP in chapter 12.

There is a lead time of one week for delivery to the north and south stores from the supply warehouse. This is caused by the order processing, picking and transport: delivery is very fast to the west stores. The despatches from the supply warehouse required to meet this are given in the lower half of the table, taking into account the week offset in the two cases. If the supply warehouse receives the goods in batches of 400, the required delivery schedule can be worked out as in the 'deliveries' line. The resultant stock at the central warehouse is also shown. Once demand at this supply echelon has been determined, the demand on manufacturing can be placed into the master schedule. Again a supply lead time has to be allowed for the distribution process.

DRP is far simpler than fair shares and it is easier to manage. An important practical difference between DRP and fair shares is that DRP relies on lot-sized quantities with prescheduled despatch times. In DRP there is normally a resultant stock caused by the batch sizes, as illustrated in Table 13.4.

Some pitfalls which have to be avoided when using DRP rather than fair shares are:

- DRP is less flexible.
- DRP focuses on each site rather than the total system.
- The sole concern of DRP is the balance of forecast and available stock, with little reference to customer service.
- DRP fixes and depends on the supply performance.

The technique is relatively rigid, since the quantities must be sent on the days scheduled. In this sense it is limited in value for transportation planners (transport must be made to fit the schedule). With DRP, quantities shown are planned, with the implication that stock is not usually currently available. Consequently, changes to shipment schedules have to be long term rather than immediate. There is little possibility of responding to shortages since the quantities shown are planned to be available, not currently available. DRP is best suited to situations of relatively steady demand: it is most successful where fixed delivery quantities are physically necessary (such as container loads).

Fair shares can be made to work in conjunction with DRP by making the best allocation of the shipments arriving into the network.

Wherever possible, the manufacturing and distribution systems should be synchronised, thus reducing the amount of inventory carried throughout the system. This could be achieved in the example shown in Table 13.4.

The interface between the computed lot-size requirements and the master schedule is the gross requirement from the supply warehouse, taking into account any finished goods stock held by the manufacturer. In practice, the DRP gross requirement can be exceeded by the master schedule and products can be planned in advance of the DRP schedule dates if it is beneficial for manufacturing planning and efficiency.

The distribution quantities calculated by the forecasting techniques and apportioned by fair shares often have to be adjusted because of other influences. The total shipment may have to be phased because of space limitation in a particular stores. The distribution quantities could well be rounded to the nearest pack size or altered to take advantage of a distributor's cost structure.

Traditionally forecast demand would be made at the product level. DRP allows the forecast to be conducted at the SKU level. A consequence of this is the need to aggregate the local forecasts to generate forecast demand for each product and ultimately product families. It will have been observed from the example that DRP results in considerable unevenness of the gross requirements. This is caused by the fixed lot sizes. If the lot size were smaller, this effect would be minimised but the number of deliveries would be increased.

Review

The development of inventory management is a dynamic activity, with new approaches being made, techniques being refined and new challenges being met. Successful inventory management is a balance between the use of basic techniques which are easy to use and give reasonable results and sophisticated techniques which can give better control if they are used properly. Modern inventory managers have the opportunity to use better and more effective controls.

It is imperative that the challenges are met, since continuing competitiveness is based on ever-improving customer service and ever-reducing inventory costs. The approaches discussed in this book give companies major improvements in their efficiency across all types of business. Understanding the full implications and applications is a gradual process, but worth the effort because of the tremendous benefits.

Appendix

Questions and answers

Questions

1 What are the three key objectives of inventory control?

2 Which departments put pressure on Inventory Managers, and what are their objectives?

3 'More stock gives better availability.' Challenge this conventional view.

4 Two companies distribute fast moving consumer goods. Their results for last year were:

Annual accounts data	Ace Distributors, $million	Deuce Associates, $million
Sales turnover	8	26
Fixed assets	2	5
Cost of sales	6.2	21
Stock value	4	8

You have been offered the job of inventory manager by both companies. Which one would you take and, based on the data provided, explain why.

5 The aim for Ace Distributors is to reduce the stock to $1.5 million. What will be the return on assets then?

6 What change in the return on capital (as a percentage) will there be when a company has reduced its stock value by $500 000, given the following data:

$$\text{Return on capital} = \frac{\text{Annual Profit}}{\text{Total capital employed}}$$

Where capital employed = stock + capital plant

Sales turnover	= $3 million
Profit	= $400 000
Capital plant	= $1.1 million
Stock before reduction	= $1.5 million

7   A company has decided to provide faster service to its ultimate customers. The 4 hour response to customer demand must now become 2 hours. This can be accomplished through the purchase of 12 local stores and locating stock at these. At the same time, the availability ex-stock at the central warehouse is to be increased from 85 to 95% across all items. The supply lead to the central stores from suppliers is currently 4 weeks and the proposal is to transfer between central warehouse and local stores weekly as required.

The following data is available:

Class	% of items	Average usage rate per week per item	MAD	Value, $ Each
A	10	100	10	65
B	20	50	20	25
C	70	4.76	1	30

(You can assume that all items in each class have the same characteristics.)

a)   How should the activities be organised?
b)   Determine the effect of the changes on inventory value.

8   How does a company give good customer service?

A company is distributing a range of fast moving items to stockists. The following is a list of despatches for one day. What is their customer service level, assuming this to be a normal day?

Order No. 1A

	Quantity requested by customer	Quantity despatched from stores
Item No. 1	20	15
Item No. 2	100	100
Item No. 3	50	20

Order No. 2B

	Quantity requested by customer	Quantity despatched from stores
Item No. 1	60	40
Item No. 2	150	100
Item No. 3	20	20

Order No. 3C

	Quantity requested by customer	Quantity despatched from stores
Item No. 1	10	10
Item No. 2	70	70
Item No. 3	50	20
Item No. 4	25	25

9   Which items in the following list are in the A, B and C classes? How do you determine this?

Item code	Weekly demand	Unit price
A901	2	$3.00
B662	2	$8.00
C355	20	$4.00
D523	40	$0.50
E191	10	$65.00
F807	1	$9.00
G010	3	$1.00
H244	1	$170.00
J459	10	$3.50
K488	22	$0.50

10   If the current stock of the items in question 9 were as shown below,

Item	Stock quantity
A901	6
B662	10
C355	50
D523	60
E191	50
F807	50

Item	Stock quantity
G010	12
H244	5
J459	20
K488	5

(a)   What is the stock cover for each of the items?
(b)   What would be the first task resulting from this analysis?
(c)   What is the stock cover value for the total stock?

11   A stores is offering ex-stock delivery to a wide variety of customers. Item 35721 is a non-seasonal item which has the following usage characteristics:

Weekly usage: 200
MAD   50
Supply lead time: 4 weeks

a)   If deliveries are brought in every week, what is the target stock level if 80% availability ex-stock is required?
b)   If the supplier is persuaded to deliver daily, what would be the new target stock level?
c)   How would the change affect the value of stockholding?

12   The statistics on the stock of a range of items is as follows:

Item No.	Unit cost	Annual usage
2A32	$25.00	12
2B44	50¢	360
3D10	10¢	120
3E82	$40.00	1
3F66	$2.00	40
4G19	$5.00	520
4H95	$4.00	7
5J53	$20.00	5
7N78	30¢	200
9P21	$120.00	5

The stores controller is under pressure from the management to reduce the stock level to one month's usage. The stores personnel are adamant that they are not going to do any extra work. Show what the stock policy should be in these conditions and how the target is achieved. (Assume that safety stock is not necessary in this case.)

13   Stock levels are as follows:

A items	$28 000	130 items
B items	$6 500	350 items
C items	$3 500	600 items
Obsolete items	$2 000	200 items

Within the budget for next year, the estimated usage rate for each is:

A items	$133 000
B items	$16 000
C items	$2 000
Obsolete items	Nil

What is the best way to reduce the stock by $2000?

14   Do you believe that JIT simply passes the burden of holding stock on to the supplier?

15   Your Managing Director has asked for a short report on ways of reducing work-in-progress and throughput time. At present, batch sizes are about 15, set-ups average 3 hours, processing time per unit is 2 minutes, and WIP stands at 13 weeks. Create a short report outlining in simple language the measures you would take.

16   a)   What are the financial benefits to a company through setting up a linear flow Kanban system?
       b)   A company is intending to set up a Kanban system for production. Comment on the changes which will be required in:
              i)   Responsibility
              ii)   Layout
              iii) Operator training
              iv)   Quality management

17   What is the role of safety stock, and why is it often not required in a JIT environment?

18   'JIT is a good theoretical idea but cannot work in our company' said the production director. 'We have customers who change their minds, a

variety of processes, many products and real people who cannot always be relied upon.'

Set out the arguments which you would make to persuade this director that JIT is a good idea for the company.

19   The data below is the stock record for item 345.

Date	Receipts	Unit cost	Issues	Balance
14/3	10	11		10
12/4			3	7
22/5			2	5
10/6	5	15		10
30/6			3	7
17/7			2	5

a)   What is the value of the last issue (17/7) of item 345, based on:
   i)   FIFO costing
   ii)   Average costing
   iii) LIFO costing
b)   If you needed to reduce the stock value in your stores by $500 000, what items would you look at first and what techniques would you use to reduce the stock value?

20   Why is safety stock held?

21   What are the three factors which determine safety stock?

22   The usage for two items over the past 5 weeks has been:

Item	Week 1	Week 2	Week 3	Week 4	Week 5
D523	60	30	20	40	50
K488	30	15	0	65	0

What are the values of average demand and variability for each?

The items cost 50¢ each and the lead time from the supplier is 2 weeks. The inventory controller is deciding whether to try and provide a service level of 90 or 95%. What is the additional cost of the better service?

23   ABC Distributors have the UK dealership for a well known manu-facturer of pre-packaged confectionery. Their main activity is the supply of goods to the local retailers ex-stock.

At the recent stocktake, ABC Distributors had 1280 different stock lines with a total value of $40 000. They order all their items during the week they fall below the reorder level. The system works well except for the fact that the stores manager and his assistant are overloaded with work – in fact they are currently 3 weeks behind with the ordering. As they are ordering one month's requirement at a time, there are often times when the stock runs out. The suppliers take 2 months to deliver, and the shortfalls cannot therefore be rectified immediately. What are the changes in operating prac-tice which you would suggest to assist them? What controls should they use which will bring the changes into effect?

24   'Re-ordering systems are not designed to stop you running out of stock.' Give two reasons why this statement can be justified.

25   What factors do you consider when setting stock levels for an inde-pendent demand item?

26   The demand history if item XX1 is as follows:

Week	Demand
1	60
2	50
3	70
4	60
5	80
6	70

The items are supplied in boxes of 100. Work out:

a)   Review level (re-order point)
b)   Average stockholding

(You can assume a safety factor of 2 for 94% customer service.)

27   Explain why, for some A items, we can expect the review level to always be higher than the physical stock level?

28   A company is putting expensive fountain pens into boxes at a rate of about 250 per week, together with ink cartridges. The stock controller has

decided that the pens they buy are class A, while the cartridges and boxes are class C. Suggest order and delivery patterns and order quantities for the three types of item, and discuss why you have chosen them.

29 a) It has been remarked that 'stock level for an item does not depend on its lead time'. Defend this remark using examples, and discuss why the statement is not precisely true.

  b) Calculate the reorder level for the supply of umbrellas given the following data:
      Lead time = 2 months
      Customer service level expected is 90%
      Demand over the last 6 months has been 11, 20, 8, 15, 10 and 14.

  c) Discuss the assumptions which you have made in using this calculation.

30 A company is proposing to schedule in supplies of 20 key items each week instead of ordering them monthly.

a) What preparatory work do they need to do with the suppliers?

b) If the purchase value of these items is $1 million per year, by how much will the average inventory be reduced?

c) What are the benefits and extra jobs caused by scheduling?

31 What factors determine the delivery quantity?

32 Discuss ways in which stock can be reduced by suppliers and customers managing a supply chain together. What are the pre-requisites for successful supply partnerships?

33 What is the difference between a target stock level system and a re-order level system?

34 a) Weekly sales of a product are:

Week	Demand
1	12
2	14
3	8
4	16
5	13
6	18

Discuss what forecasting techniques can be used to determine the demand for the next week. Indicate which you would prefer to use and the reasons for your choice.

b)   If the initial forecast for week 1 was 10, work out the forecast demand for each week using exponential smoothing, and give your prediction for week 7. (You can use an α factor of 0.2 for this calculation).

35   What items of data (inputs) are required for the MRP calculation? For each, state the level and type of accuracy necessary to ensure reasonable results.

36   A watch manufacturer has a schedule for assembly covering 10 weeks. The schedule is:

Watch X   Manufacturing orders for 1000 watches due in weeks 2, 4, 6, 8 and 10. Assembly lead time per batch is 1 week.

Watch Y   Manufacturing orders for 500 watches due in weeks 3, 6 and 9. Assembly lead time per batch is 2 weeks due to a special processing on the watch cover.

Watch Z   Manufacturing orders for 100 watches due every week. Assembly lead time is half a week. The parts for the batch of week 1 have already been issued.

Watches X and Y use the same module A and batteries (2 per watch). Watch Z uses a different module B and the same batteries but only 1 per watch. Batteries and modules come in boxes of 500. There are 7 boxes of batteries and 3 boxes of modules in stock.

The buyer has outstanding purchasing orders for a total of 20 000 batteries and 6000 modules, half quantities to arrive in weeks 4 and 10.

a)   What would you expect your MRP package to recommend for purchasing? (Assume that the package is not capable of recommending re-scheduling).

b)   What would you recommend should be done?

37   A product is manufactured in two stages. The material for the process consists mainly of one basic, bought out ingredient. Expected sales of the product are as follows:

Week	1	2	3	4	5	6	7	8	9	10
Sales	210	300	140	65	200	260	150	100	190	220

The stock situation is:

Raw material   500
Semi-finished   220
Finished goods   650

The manufacturing and supply situation is:

Operation	Lead Time
Purchasing	3 weeks
Bulk production	1 week
Finishing	2 weeks

What should the weekly purchase plan be on a lot-for-lot system? And with a 500 batch size?

38 a) What different underlying assumptions are there between DRP and fair shares allocation of goods?
    b) What is the fair shares priority logic for distributing stocks to regional warehouses?

39 Prepare a manufacturing schedule, using the DRP technique, for the following forecast demands and lot size replenishment quantities.

	Lot Size			Period Demand					Current
		1	2	3	4	5	6	7	Stock
London	600	150	320	370	190	150	320	370	600
Manchester	250	70	70	70	70	70	70	70	230
Newcastle	400	90	150	150	90	90	150	150	310
Internal	Monthly								
Sales	Call-off	30	30	30	30	30	30	30	

Lot size for manufacturing store is 1000.
Current stock for manufacturing store is 1650.
Replenished lead time for all warehouses is two periods, internal sales are replenished immediately, manufacturing lead time is 3 weeks.

40 What are the advantages and disadvantages of keeping stock in stores rather than in depots.

41 The following 10 items are the range of stock held in a small stores.

Item No.	Unit cost ($)	Annual usage
1	150	2
2	125	100
3	85	10
4	50	70
5	40	1
6	26	20
7	20	3
8	10	1
9	7	300
10	6	20

a) Identify which are the A, B and C Class items.
b) Discuss how ABC is used in purchase ordering.
c) Review another use of this Pareto classification.

42 The following data is available for the demand for an item:

Week	1	2	3	4	5
Demand	10	20	5	10	15

What would you estimate to be the average quantity which is held in stock to meet the demand for this item, treating it as:

a) Independent demand
    (Assuming that a 99.2% customer service level is required).
b) Dependent MRP demand (level 2)
    (Assuming a one week lead time)
c) JIT supply
    (Assuming a one day delivery interval)

Specify any assumptions you make.

Answers

1 See text, Chapter 1.

2 See text, Chapter 1.

3 High stocks result from poor control, and poor service results from poor management of supply and forecasts. The better the inventory management, the better the availability and the less stock cover is required.

4

Annual accounts data		Ace Distributors, $million	Deuce Associates, $million
Annual profit		1.8	5
Return on sales		1.8	5
		8	26
	=	22.5%	19.2%
Assets are		2 + 4	2 + 4
Return on assets		1.8	5
		6	13
	=	30.0%	38.5%

Ace has better margin and more potential for improvement in stock control.

5 51.4%

Annual accounts data		Ace Distributors, $million
Annual profit		1.8
Return on sales		1.8
Assets are		2 + 1.5
Return on assets		1.8
		3.5
	=	51.4%

6

Capital employed before reduction = Stock + Capital Plant
= $1.5 million + $1.1 million
= $2.6 million
Capital employed after reduction = $2.6 million − $500 000
= $2.1 million

Return on capital before reduction = $400 000/$2 600 000
                                                                    = 15%
Return on capital after reduction   = $400 000/$2 100 000
                                                                    = 19%

7   (a)   Key features of the organisation should be:
            Integrated inventory system
            Central control of inventory
            Enclosed and controlled local stores
            (see Chapter 13)
    (b)   Increase in service level will increase the safety stock
            CSF for 85% customer service is   1.3 in MADs
            CSF for 95% customer service is   2.06 in MADs
            The increase in stock required to give
            the improved service is therefore   2.06/1.3 = 58%

The stock required in each of the 12 stores will be enough to cover between the deliveries (i.e. 1 week's usage) plus safety stock. If the total usage rate is 'D', then the usage at each store is D/12. As the MAD only reduces as the square root of the volume then the average stock at each local store will be:

$$\text{Average local stock} = \text{Half weekly usage} + \text{Safety stock (Half}$$
$$D/12) + \text{CSF} \times \text{MAD/(Square root of 12)}$$
$$\times 1\ D/24 + 0.38*\text{MAD}$$

Assuming that availability at local stores is still 85%

$$12 \text{ local stocks} = 0.5*D + 4.5*\text{MAD}$$

$$\text{Average central stock} = D + 2.06*\text{MAD}*(\text{square root of 4})$$
$$D + 4.12*\text{MAD}$$

$$\text{Total stock} = \text{Average central} + \text{Average total local}$$
$$= D + 4.12*\text{MAD} + 0.5D + 4.5*\text{MAD}$$
$$= 1.5D + 8.62*\text{MAD}$$

Calculation of new stock value:
Using ABC analysis and the figures in the question:

	Av stock/item	Total stock value
A Stock = 236.2		153 530
B Class = 247.4		123 700
C Class = 15.76		33 096
TOTAL per 100 stock lines		310 326

Original stock values was:

D/2 + 1.3*MAD*(Square root 4)

	Av stock/item	Total stock value
A Stock = 76	49 400
B Class = 77	38 500
C Class = 4.98	10 458
TOTAL per 100 stock lines	98 358

The change in inventory was therefore from \$98 358 to \$310 326 an increase of 315.5%.

8   Solving customer requirement, often by delivery on time (see text)

Demand satisfied on time:	Requested	Despatched	
By order	0	3	0%
By order line	5	10	50%
By item code			
Item 1	65	90	72%
Item 2	270	320	84%
Item 3	60	120	50%
Item 4	25	25	100%
TOTAL	420	555	
Average line availability			77%
Average item availability			76%

Different answers are used in different situations (see text)

9

Item code	Weekly demand	Unit price	Turnover	Rank	Class
A901	2	\$3.00	\$6	9	C
B662	2	\$8.00	\$16	6	C
C355	20	\$4.00	\$80	3	B
D523	40	\$0.50	\$20	5	C
E191	10	\$65.00	\$650	1	A
F807	1	\$9.00	\$9	8	C

Item code	Weekly demand	Unit price	Turnover	Rank	Class
G010	3	$1.00	$3	10	C
H244	1	$170.00	$170	2	B
J459	10	$3.50	$35	4	C
K488	22	$0.50	$11	7	C

See also text, chapter 3.

10   (a)

Item	Stock	Usage	Cover
A901	6	2	3
B662	10	2	5
C355	50	20	2.5
D523	60	40	1.5
E191	50	10	5
F807	50	1	50
G010	12	3	4
H244	5	1	5
J459	20	10	2
K488	5	22	0.23

(b)   Ensure that supplies of item K488 are arriving within 1 day.
(c)   This is calculated on a financial basis.

	Stock	Usage	Unit Value	Stock Value	Turnover
A901	6	2	$3	$18	$6
B662	10	2	$8	$80	$16
C355	50	20	$4	$200	$80
D523	60	40	$1	$30	$20
E191	50	10	$65	$3250	$650
F807	50	1	$9	$450	$9
G010	12	3	$1	$12	$3
H244	5	1	$170	$850	$170
J459	20	10	$4	$70	$35
K488	5	22	$1	$3	$11
				$4963	$1000

Stock cover for all stock = $4963/$1000
                          = 4.96 weeks

11   (a)

TSL = Demand in lead time and review period plus safety stock
        in 1T and RP

$$= 200 * (4+1) + 1.3 * 50 * \left(\text{Square root}(4+1)\right)$$
$$= 1000 + 145.3$$
$$= 1145.3$$

(b)   Assuming that the lead time is still 4 weeks and orders are placed
        daily:

$$= 200 * (4+0.2) + 1.3 * 50 * \left(\text{Square root}(4+0.2)\right)$$
$$= 840 + 133.2$$
$$= 973.2$$

(c)   This would be a reduction of 172.1 in TSL
        The average stock is TSL – Half usage in (LT+RP)
        So reduction would be
        from 645.3
        to 553.2
        Reduction 92.1 units at standard cost

12   See text, Chapter 3.

13   See text, Chapter 3.

14   See text, Chapter 4.

15   See text, Chapter 4.

16   a)   Inventory saving
            Higher efficiency
            Less scrap
            Reduced space
            Cash flow
            i)   Shop floor responsibility for quality, throughput, working
                  methods, job design.
            ii)   Linear, compact, flowline, no stores.

    iii)   Cross training on variety of equipment, empowerment

    iv)   Responsibility to operators, self checking, problem solving.

17   To resolve the mismatch between supply and demand.
     To enable fast delivery to be provided for long lead-time items.
     To avoid quality and delivery problems affecting customers.
     JIT reduces lead time and requires high level of quality so safety stock is not required.

18   Financial and cash flow.
     Customer service and efficiency
     Flexibility and empowerment
     New role for shop floor
     Quality environment

19

Date	Receipts	Unit cost	Issues	Balance	No. left 14/3	From 10/6	Average cost
14/3	10	11		10	10		11
12/4			3	7	7		11
22/5			2	5	5	5	11
10/6	5	15		10	5	5	13
30/6			3	7	2	5	13
17/7			2	5	0		13

So, value of issue of the two items was:

a)   i)   22

       ii)   26

       iii) 30

b)   See text, Chapter 5.

20   See text, Chapter 6.

21   See text, Chapter 6.

22

Item	Wk 1	Wk 2	Wk 3	Wk 4	Wk 5	Average	
D523	60	30	20	40	50	40	
K488	30	15	0	65	0	22	
						MAD	MADP
D523	20	10	20	0	10	12	30%
D488	8	7	22	43	22	20.4	93%

Lead time 2 weeks   CSF   90%   1.6
                            95%   2.06

	MAD	SS 90%	SS 95%	Increase	Extra cost	% of turnover
D523	12	27.15	34.96	7.81	3.90	20%
K488	20.4	46.16	59.43	13.27	6.64	60%

23   See text, Chapter 7.

24   See text, Chapter 7.

25   See text, Chapter 7.

26   See text, Chapter 7.

27   The reorder level is used to trigger new replenishment orders when the stock + supply orders outstanding become low. If the delivery frequency is shorter than the lead time then the stock is low, delivery quantities are small and there are always supply orders outstanding. Hence for long lead time A class items ordering is frequent and stock is always less than the review level.

28   See text, Chapter 7.

29   a)   Stock level depends mainly on supply lead time, customer service and demand pattern. Lead time variation is only as the square

root. (However, delivery performance may be lead time dependent also.)

b) Average = 13
    MAD = 3.33
    Customer service factor is 1.6
    Safety stock = $1.6 \times 3.33 \times SQRT(2) = 7.53$
    Reorder level = $13 \times 2 + 7.53 = 33.53$

c) Continuous demand
    Continuation of the pattern
    Normal distribution and non-seasonal

30 See text, Chapter 8

31 See text, Chapter 8.

32 Better forecasting.
    Treating demand as dependent and avoiding safety stock effects.
    Measuring end user usage, structuring supply source.
    Integrating systems
    Square root of warehouses effect.

33 See text, Chapter 8.

34 (a) Extrinsic forecasts to give causes of demand
         Market analysis and surveys
         Historical techniques

Moving average	OK as demand is random and varying much
Exponential smoothing	Better since there could be a slight trend
Double exponential	Will be over-reactive until a trend is seen
Base series	Not appropriate as seasonality not established
Regression	No advantage over other methods
More sophisticated techniques	Insufficient data for these

(For details see text in Chapter 10)

(b) Exponential smoothing with $\alpha = 0.2$

	Week	Demand	Forecast
	1	12	10
	2	14	10.4
	3	8	10.8
	4	16	9.6
	5	13	11.2
	6	18	10.6
Forecast	7		11.6

35 See text, Chapter 12.

36 See text, Chapter 12.

37 See text, Chapter 12.

38 See text, Chapters 12 and 13.

39 See text, Chapters 12 and 13.

40 See text, Chapter 13.

41

Item No.	Unit cost	Annual usage	Turnover	% turnover	Cumulative % turnover
2	125	100	12 500	62.5	62.5
4	50	70	3500	17.5	80.0
9	7	300	2100	10.5	90.5
3	85	10	850	4.3	94.8
6	26	20	520	2.6	97.4
1	150	2	300	1.5	98.9
10	6	20	120	0.6	99.5
7	20	3	60	0.3	99.8
5	40	1	40	0.2	100.0
8	10	1	10	0.1	100.0
Total		527	20 000		

41 (a) A Class   Item 2
        B Class   Items 4, 9
        C Class   Items 3, 6, 1, 10, 7, 5, 8
    (b) ABC determines delivery quantity and reduces purchase work-
         load (see Chapter 3)
    (c) Discussion of optimising profits, lead time reduction or cycle
         counting.

42 a)   Independent demand
         Assume that a 99.2% customer service level is required.

         Average demand = 12
         MAD   = 4.4
         Customer service factor   3
         Safety stock   = 3 × 4.4 × square root (1)
                             = 13.4
         Delivery quantity = 12
         Average stock   = Half delivery quantity + safety stock
                             = 6 + 13.4
                             = 19.4

    b)   Dependent MRP demand (level 2)
         Assuming no safety stock is required, stock would be:

Week	0	1	2	3	4	5
Demand	10	20	5	10	15	

         Average stock = 12

    c)   JIT supply
         Assume that demand is the same every day. (Alternative assump-
         tions are that there should be a safety stock calculated as (a) or
         as one day lead time.)

Week	0	1	2	3	4	5
Demand	10	20	5	10	15	
Daily demand	2	4	1	2	3	

         Assuming that demand is despatched every day:
         Average stock = half day's demand:
                            1   2   0.5   1   1.5
         So average stock during the period is 1.2 (assuming a one day
         delivery interval).

Index